Home Plumbing

THE COMPLETE HANDBOOK

JULIAN BRIDGEWATER

NEW HOLLAND

This edition first published in 2008 by New Holland Publishers (UK) Ltd
Garfield House, 86 Edgware Road
London W2 2EA
United Kingdom
London • Cape Town • Sydney • Auckland
www.newhollandpublishers.com

ISBN 978 184773 339 9

Editorial Direction: Rosemary Wilkinson
Editors: Gareth Jones and Fiona Corbridge
Design: AG&G Books
Illustrations: Peters & Zabransky (UK) Ltd
Cover photograph: Ian Parsons

Printed and bound by Kyodo Nation Printing Services Co.,Ltd

NOTE
The author and publishers have made every effort to ensure that all instructions
given in this book are safe and accurate, but they cannot accept liability for any
resulting injuries or loss or damage to either property or person, whether direct
or consequential and howsoever arising.

Contents

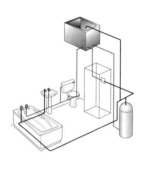

Introduction

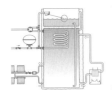

New homes, old homes, townhouses, bungalows, cottages and cabins all need an efficient plumbing system. There is clean, fresh drinking water to be piped in, hot and cold water to be piped to kitchens and bathrooms, wastewater to be piped out, and rainwater to be piped down from the roof. And of course, where there is a plumbing system, there are pipes, cisterns, tanks, drains, U-bends, taps, washers, valves, stopcocks, ballcocks and so on. The downside of a sophisticated plumbing system like this is that sooner or later you are going to need the skills of a plumber to service, repair, replace, enlarge and update it.

Most people prefer not to think about their plumbing until they are confronted with frozen pipes,

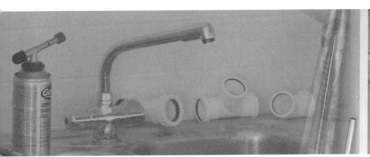

a burst storage tank, blocked drains, a leaky cistern or some other problem. Plumbing seems to be one of those activities that fills householders with dread. Perhaps it is an echo of times past, when a source of clean, fresh water was everything. If the water in the well or stream dried up, the very viability of a home was called into question. Whole villages were abandoned simply because the water source ran out. Or perhaps the popular reputation of plumbers is at fault – plumbing is still considered to be something of a black art, with plumbers' main talents residing not in pipe-bending, but time-and-money-bending. But *Home Plumbing* will rescue you. We tell you all you need to know – from how to select a good plumber to how to roll up your sleeves and do it yourself.

Best of luck!

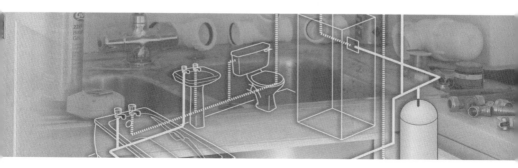

CHAPTER 1

Preliminaries

What does your home need?

When plumbing goes wrong, it has the potential to go severely wrong! Take the example of the house where the owners went on holiday for a week over Christmas, and a burst pipe sprayed water all over the loft. The whole house became one big icicle, with just about every room ruined – plaster falling off the walls, doors warped and window frames bowed – everything was a write-off.

Is your plumbing a wet and watery time bomb just waiting to explode, or does it simply need updating? Is the subject of plumbing simply something that you haven't thought about? If the answer to any of these questions is 'yes', it's time to confront the issue. Take a notepad and pencil, and have a good look at the entire plumbing system – inspect the loft, cellar, kitchen, bathroom, toilet, outdoor pipes – everywhere. Make a hitlist of problems that must be resolved, and a wish list of changes that you might like to make if you could afford to.

Hitlists and wish lists

The hitlist should consist of annoyances that need tackling fairly quickly, such as a leaky tap, a bath plug that doesn't really fit, an overflow that drips, a copper pipe that looks to be particularly green, a drain that oozes, and so on.

For the wish list, put together a programme of changes that can be tackled as time and budget allow. You might like to move the kitchen sink to the other side of the room to take advantage of a better view as you wash up, or to jazz up the bathroom with new taps and a power shower. The needs of the household may be better served by relocating the bathroom, or even adding an extra bathroom. Do you want to remove an outside toilet? Do you want to fit a tap in the garage? Do you want bigger radiators or to move existing radiators?

Restrictions

There are certain areas of the plumbing system that you cannot tamper with, such as the stopcock by your gate and the underground pipe between that and the water supply in the road, both of which belong to the water supply company. If by chance you damage either of these, you must report the problem to the supplier. You must never intentionally make alterations to this stopcock. Tell the supplier if the tap is corroded or damaged (for example if it leaks, or you cannot turn it off).

There are strict regulations regarding the installation and servicing of gas boilers. You are allowed to change and service the water pipes in and around the boiler (to the pump and suchlike), but you cannot touch the gas supply pipe. If you have any doubts about work you have done in the past, or the work that you are doing at the moment, have it checked by a CORGI engineer (registered by the Council for Registered Gas Installers).

There are many new health and safety regulations concerning the

10 Expert Points

BEFORE YOU MAKE ANY CHANGES TO THE PLUMBING, FOLLOW THESE TEN STEPS TO GET TO KNOW YOUR PLUMBING SYSTEM.

1 WATER SUPPLIER'S STOPCOCK
Look for the water supplier's main stopcock outside the house. This will usually be in the pavement by the front gate, or sometimes in the garden near the road (or sometimes you will find both). It will be below ground level, usually under a hinged cast-iron flap. Make sure that it can be turned off easily.

2 MAIN INTERIOR STOPCOCK
Find out where the main stopcock is inside the house. It is usually located under the kitchen sink, but could be just about anywhere. Make sure that it works.

3 INTERIOR PIPES
Assess the condition of the pipes inside the house – especially the cold-water feed (it will feel cold to the touch). Look under sinks and baths, in the loft, in cupboards, in the garage, and so on. Have any of the pipes acquired a green tinge? If so, they may be nearing the end of their life.

4 FIND OUT HOW OLD THE PLUMBING IS
To head off potential problems, the plumbing needs inspecting if it is more than about ten years old.

5 LEAD PIPES
If you find lead pipes in your plumbing system, they need to be replaced. Your local authority will give you a grant towards the cost of doing this work.

6 TAPS
Taps are good indicators of the age and state of the plumbing. If a tap won't stop dripping, and if the constant dripping is resulting in brown water stains down the side of the bath and the washbasin, then not only do the washers need replacing, but there is a good chance that the body of the tap itself is damaged. If this is the case, it needs to be replaced.

7 TOILETS
Look behind the toilet, where the large pipe goes down through the floor or into the wall. If there are water stains on the wall or floor, the seals need replacing.

8 COLD-WATER STORAGE TANK
Go up into the loft and inspect the cold-water storage tank. Old tanks made of lead, galvanized steel or asbestos will need to be replaced as a priority.

9 BALLCOCKS
While you are in the loft, have a look at the ballcock float valve inside the cold-water storage tank. If the ball is dented and green, it is made of copper. Copper floats have been obsolete for a good number of years, so if you have one it is old and may be leaking, and will therefore need to be replaced as soon as possible.

10 ELECTRICS
Look for electric bonding links (green and yellow wires that link large metal items, such as a bath, to copper plumbing pipes). Make sure that these links are continuous. (For further information, see 'Cross bonding' in the 'Glossary of terms, tips and materials' on page 169.)

point at which the water supplier's clean water meets your taps and valves – for example the garden tap, and the cold-water storage tank and its valve.

If you are considering a DIY task that involves cold-water feed, make contact with your water supplier and ask for their advice before you begin.

Locating your pipe runs

It's vital that you know the location of all the water pipes on your property – in walls, under floors, in the loft, out in the garden – because it's the easiest thing in the world to accidentally puncture a water pipe. For example, a relative was banging a nail into a floorboard as part of a minor weekend task and suddenly, before he could say 'Niagara Falls', he was trying to stop a fountain of water spurting up from the floor. And when our neighbour decided to bang an iron stake into the ground late at night – no, we didn't ask what he was doing – he damaged an underground pipe. One moment he was only ten minutes away from a good night's sleep, and the next he was running around in the dark desperately trying to turn off the water. The moral of these stories is to know where your water pipes are.

How old is your home?

The age of your home will, to a great extent, be responsible for the character of your plumbing problems. For example, in one house we lived in, built in about 1860, all the pipes and even the storage tanks were made of solid lead. Knowing that lead is toxic, we contacted the council for advice and ended up with a full grant towards the cost of a complete plumbing replacement. We sold the lead pipes at a nice profit, and used the lead tanks as decorative features in the garden. In another house, the pipes were old, pre-metric sizes. When we attempted to join new pipes to the system, nothing fitted properly.

In a house in Somerset, built in 1930, the main cold-water feed pipe (the rising main) that ran from the under-sink stopcock to the cold-water tank in the loft was so decrepit that when we increased the water pressure by turning the stopcock on, the pipe started to weep at various weak points. We also lived in a bungalow where the hot-water pipes had been channelled into the concrete floors and skimmed over with cement. The problem was that the pipes were only a few millimetres below the surface. When we attempted to lay a carpet, we accidentally spiked one of the pipes. As if that wasn't bad enough, we had just moved in and had no idea where the stopcock was located.

Yet another home – an old Victorian semi – brought more plumbing delights. We were puzzled because there didn't appear to be a drain for the wastewater that ran from the kitchen sink. A close inspection revealed that the previous owner had solved the problem of a blocked drain by bending the lead waste pipe back on itself so that it disappeared through a broken airbrick in the wall of the house. We lifted a floorboard to discover that the wastewater had been draining away around the footings of the house, and this had resulted in some very interesting football-sized growths of fungus.

So if your house is more than 30 years old, start by mapping out the position of the various pipes and shut-off valves. If the house is more than 50 years old, and there is a chance that there are lead pipes, ancient cesspits or other undesirables, it is best to call in a specialist to do a survey.

To recap: be on your guard and expect the worst if you encounter things such as lead pipes and tanks, asbestos tanks and gutters, cesspits or septic tanks leaking ooze, or a confusing mixture of lead, galvanized steel and plastic pipes. Prepare for the unexpected by finding out how to turn off the water supply.

Type of structure

The age, type of construction and form of your home will have an impact on the level of difficulty involved in making changes to the plumbing. For example, if pipes are buried under concrete floors and start leaking, you either have to dig up the floors and replace the pipes, or leave the pipes where they are and install a modern system that runs above floor level. In our home, old galvanized steel central-heating pipes were buried in concrete floors, so we simply cut them off, left them in place and bypassed them with new copper pipes. In one or two farmhouse-type homes that I have worked on, the main feed pipes from the stopcock to the cold-water tank in the loft were made from blue polypropylene. This type of system isn't very attractive and has huge joints, but it doesn't matter if there is room for it to be hidden away in the structure of the house.

Old houses

In very old houses, built at a time when the best that you could expect in the way of plumbing was a scullery or wet room with a hand pump over a trough sink, the plumbing is often a horrible mish-mash of pipes installed over more than 100 years. There might be lead pipes that were installed in the 1890s, galvanized steel pipes from the 1940s, imperial-sized copper pipes from the 1950s, mini-bore central-heating pipes from the 1970s, and perhaps a plastic pipe system from the 1990s. You could find lead cold-water pipes in the garden running to the kitchen sink, one or two stainless-steel pipes in the bathroom, galvanized steel pipes buried under the drive, cast-iron pipes for the drains, plastic pipes for the cold-water feed, an underfloor heating system made from galvanized steel, and so on. Such a system is fine if it works, but the problem comes when you need to make changes. Certainly you can continue the long tradition of bodge-on-bodge, but it is best to get rid of the lead and anything else that is over 20 years old and start afresh. Never leave the old pipes and tanks in the loft, because they will just add to the confusion.

Make notes

If you are at all unsure about the layout of your plumbing system, make a drawing and label the various elements. Identify the pipes, taps (hot- and cold-water feed), the water supplier's exterior stopcock to cut off the water to the house, the main interior stopcock, the stopcock from the cold-water storage tank, and so on, so that you have a clear picture of what goes where and how. Along the

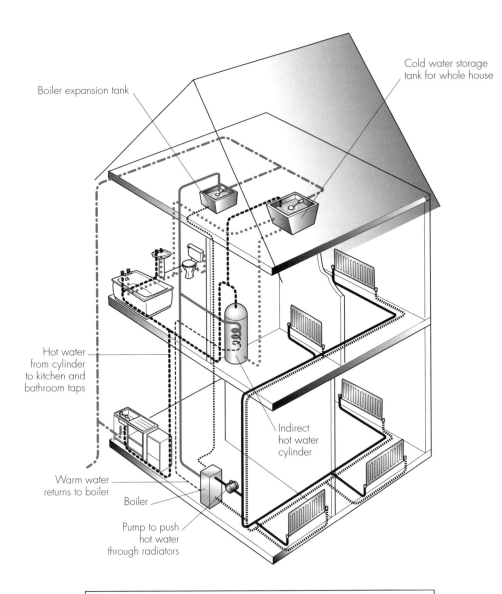

Cold water storage tank for whole house

Boiler expansion tank

Hot water from cylinder to kitchen and bathroom taps

Indirect hot water cylinder

Warm water returns to boiler

Boiler

Pump to push hot water through radiators

━━ Hot water feed to radiators	━■━ Rising supply main (cold)
▪▪▪ Cooled water return from radiators	▪▪▪ Cold supply from storage cistern
--- Cooled water from heat exchanger	●●● Hot water supply from cylinder
━━ Hot water feed to heat exchanger	⋯⋯ Boiler cold water top-up

A typical house with a traditional plumbing system.

way – when you are looking at pipes and taps, and creeping through the loft – keep your eyes open for potential problems and areas where you might make changes.

Urgent repairs

The biggest emergency you are likely to face is a water leak. There is nothing quite so upsetting as water dripping down through the ceiling, trickling through walls, or gushing up through the floor. The first step is to turn off the main interior stopcock, then decide whether to call a plumber or to tackle it yourself. If you are going to have a go yourself, you need to have an emergency repair kit at the ready. It should contain a selection of spanners, a hacksaw, a pipe cutter, a selection of compression joints, various short lengths of copper pipe in a range of diameters, half a dozen jubilee clips, an epoxy putty repair kit, a gas torch, a small pot of flux, a roll of lead solder, a pack of fine-grade wire wool, a roll of PTFE tape, and a good wind-up torch.

What to do in a catastrophic emergency

A catastrophic emergency is when there is a large water leak, and the volume of water is such that it is likely to damage the building. (See pages 146–161 for more information on how to deal with emergencies.)

● **Burst pipe**
It is a freezing cold winter's night and you have just come home from a party. The floor is awash and water is pouring down the kitchen wall.

1 Go around the house flushing the toilets and turning on all the taps. While the cold-water taps are running, fill up as many containers as possible – saucepans, kettle, bath – so that you have enough water for washing, cooking and drinking.

2 Go to the main stopcock under the sink and turn it off.

3 When the water has stopped running, take a torch and look in the loft to identify the burst pipe, rub the area around the burst with wire wool and make a temporary mend with epoxy putty, following the instructions of the manufacturer.

4 Mop up the water.

5 In the morning, either call in a plumber, or roll up your sleeves and replace the section of pipe. If the leak was due to an aged plumbing system, consider upgrading the whole system. If the problem was caused by an icy draught, it might be best to re-insulate the pipes.

● **Punctured pipe**
You have been working on a concrete floor, hammering in a nail or perhaps banging in the point of a chisel, and have hit a pipe. The stopcocks are so old that they cannot be completely turned off and there is a continual slight dribble of water running through the pipe.

1 Drop a towel over the leak so that the water doesn't spurt up and damage the walls.

2 Go around the house flushing the toilets and turning on all the taps. While the cold-water taps are running, fill up as many containers

as possible – saucepans, kettle, bath – so that you have enough water for washing, cooking and drinking.

3 Go to the main stopcock under the sink and turn it off. When the water has stopped running, mop up the mess that has been made.

4 Use a hammer and cold chisel to excavate the concrete from in and around the hole, so that there is enough room to use a small hacksaw or, better still, to spin a pipe cutter.

5 Having first made sure that you have two compression joints and replacement pipe of the right size, cut a 200 mm section out of the pipe, centred on the damaged piece. Use fine-grade wire wool to clean up the damaged ends. (Note that the advanced age of the stopcock that is dribbling water means that a soldered mend is not possible.)

6 By trial and error, cut and fit the two compression joints and a length of new pipe to bridge the gap. Use a fine rod file to remove the thin 'stop' ring at the centre of the joint so that you can slide the body of the joint along the new and old pipes.

7 One end at a time, slide a cap-nut on to the end of the new piece of pipe, followed by an olive (which slides under each cap-nut, grips the pipe and forms a watertight seal) and then the body of the joint. Tighten up with a pair of spanners so that the olive grips the pipe, and then undo the joint and remove the body.

8 Slide all the component parts into place (cap-nuts, olives, the body of the joints on the old pipe and the piece of new pipe complete with its cap-nuts and olives). Wrap a few turns of PTFE tape over each of the olives, and then re-assemble the whole thing and tighten up.

9 Finally, clean up the mess, turn on the main stopcock, wait until water runs out of the various taps, and then turn the taps off and wait until the system fills up. Keep a watchful eye on the mend – if it weeps, give the cap-nuts another quarter-turn.

New solutions

The above steps describe how to complete a traditional repair, but there are now also various quick-fix solutions available, designed specifically for the hole-in-a-pipe scenario. There are push-fit plastic joints, flexible joints, special compression joints, and so on. While many of these items are pretty good, most of them require ideal working conditions and would not be good in a situation where old pipes are dented, gritty or wet.

More often than not, an old plumbing system is characterized by ancient stopcocks that cannot be turned off completely. The steady weeping of water means that you cannot easily make a soldered repair and, in this instance, compression joints are a good option.

A new system

There comes a point in the lifetime of a plumbing layout when repairs and problems are so frequent that it is best to dig into your wallet and pay for a completely new system. Has that time now arrived? And, of course, if you

10 Expert Points

THIS TEN-POINT SURVEY WILL HELP YOU ASSESS YOUR PLUMBING SYSTEM. IF IT FAILS ON MORE THAN THREE POINTS, IT NEEDS A THOROUGH OVERHAUL.

1 WATER SUPPLIER'S STOPCOCK
Look at the main stopcock, located under a flap near your garden gate. The tap at the bottom of the hole should be relatively easy to turn on and off. If it is very stiff or doesn't turn, call the water supply company and ask them to make repairs.

2 MAIN INTERIOR STOPCOCK
Check the stopcock under the sink. Can you turn it on and off? Does the cold water stop flowing when it is in the 'off' position? If it doesn't turn, and/or it doesn't cut the water off when it is in the 'off' position, it must be replaced. This stopcock must be in first-class condition so that, in an emergency, you can turn off the water immediately.

3 TAPS
Are any of the taps dripping? If there are stains down the side of the bath or sink, you need to fit new washers. If on-view items such as taps have been neglected, there is a good chance that the rest of the system will be in a bad way.

4 PIPES
Look at the copper pipes that run around the house – under the sink, in cupboards, and so on. If they are stained green with little dots of dampness it's about time that they were replaced.

5 DRAINS
Go outside and look at the drains that take the wastewater from the bathroom and kitchen. If they are pottery with cast-iron covers, and the waste pipes are made from copper or steel, this is a good indication that they are old and need overhauling.

6 INSPECTION COVERS
Look outside for large iron inspection or manhole covers, which might be anywhere – outside the kitchen window, in the drive or hidden away under a plant container. Lift them up. The gullies, holes and traps at the bottom of the manholes should be clean, with no lumps of sludge, unpleasant smells or crumbling concrete.

7 CESSPITS AND SEPTIC TANKS
If you have a cesspit/cesspool or a septic tank, check that the whole area around the inspection covers is clean. If you notice an unpleasant smell, or the ground is soft and oozing, refer to page 63.

8 TOILETS
Remove the lid of the WC cistern. Operate the flush and watch as the system empties and fills. If the flush doesn't work first time, the action is juddery or noisy, or the water rises above the level of the overflow, it needs an overhaul.

9 COLD-WATER STORAGE TANK
Have a good look at the cold-water storage tank in the roof space. It should have a close-fitting lid. Get someone to run off some water while you observe the tank. If all is well, the water level will go down, the float will fall, and water will gush from the feed pipe. Turn off the tap. When the level of the water rises in the tank, the float should rise and the water should cut off. The level of the water should be well below the level of the overflow pipe. If the valve is dripping, noisy or intermittent, it must be replaced.

10 FLOORS
Walk around the house and look at the floors behind toilets, under sinks, in the hot-water cylinder cupboard, under radiators, and so on. If there are damp patches or water stains, have a closer look and see if you can identify the problem.

have just moved house, then the age and condition of the plumbing system will be a mystery to you. If you are unsure of the state of your plumbing, run through the basic ten-point survey on page 13. It won't provide all the answers, but it will tell you if the system is failing.

Gradual replacement

If you live in a new house and part of the plumbing fails – such as a tap or valve – it is generally safe to assume that it is only the component that has failed, rather than the whole system. But if you live in an older house, it's not such a clear-cut situation. One problem with houses built before the 1950s is that the various plumbing systems – cold water, hot water, bathroom, central heating and so on – were generally added piecemeal in response to changing needs and expectations, and in line with improved technology. First there was a pump in the yard, then there was a pump in an outhouse scullery, then there was a washroom downstairs and a toilet outside – still no hot water – then the luxury of a bathroom and toilet upstairs, and so on. These days, there is often more than one toilet in a house and there may also be an extra bathroom or shower room. All of these extra facilities will overload an already creaking old system.

The various improvements might have been put in place about every 15 years over a period of 100 years, and the likelihood is that they are going to break down in much the same order as they were built. To

avoid the expense of replacing all the plumbing in one go, a good option (especially if the more recent systems are in good condition) is to choose a regular programme of part-replacement, making renovations about every ten years.

Fitting major items

Larger plumbing tasks include the fitting of new facilities such as an instant shower or garden tap, or running the plumbing into an extension, or moving the radiators from one location to another. Maybe it is necessary to replace a bath, WC pan and cistern, or the cold-water storage tank in the loft, and so on.

Most major additions to the existing system will require you to draw up plans and work out costs, turn off the stopcocks, drain part of the system, and put in pipes and valves. Any new item needs a new pipe run complete with valves, so plan a route from the nearest existing pipework to the item. For the most part, the pipes are going to be fixed in place – either running down walls from ceiling to floor, or horizontally at skirting level – and this will necessitate a certain amount of drilling into plaster and maybe cutting into wood. There is bound to be some mess! It is possible to run pipes behind existing skirtings, and/or to cover them with panelling, but most householders prefer to have the pipes on view. If you are running pipes to an outside workshop or garage, make sure that they are protected from damage and insulated against the cold, and do not present a hazard.

14

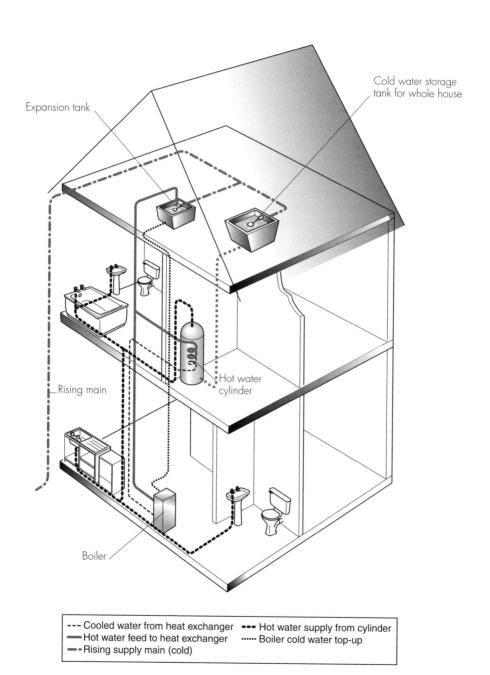

Cold water storage
tank for whole house

Expansion tank

Hot water
cylinder

Rising main

Boiler

--- Cooled water from heat exchanger   ••• Hot water supply from cylinder
— Hot water feed to heat exchanger   ••••• Boiler cold water top-up
–•–Rising supply main (cold)

A basic hot water supply system.

CHAPTER 2

Employing a plumber

Plan of action

Decide whether to do the work yourself or call in a plumber. If you would (at least in part) like to go down the DIY road, read though this book carefully and search out all references that relate to your project. Look at the 'how to' sections and try to balance the various tasks against your skill level. For example, while you might not feel up to re-plumbing the whole house, or even fitting a new bathroom, you might feel confident enough to change all the washers or maybe to fit new taps.

Look at the tasks and decide how long they are going to take to complete. Will you have enough time? Are they tasks that you can do over a weekend, or will you need to do them during a holiday? Would it be best to do the job in the summer?

You must be very clear about your aims and needs. List precisely what you want to do and either use it as a personal guide, or show it to your chosen plumber so you can discuss the various options. As with everything else in life, plumbing items and materials can be bought in various grades and qualities. For example, a new bathtub could be made of cheap plastic, pressed steel with an enamel finish, or cast iron.

Finding a good plumber

At some point you will need the services of a qualified plumber –
either to check the standard of your DIY efforts, or to do a job that is too much of a challenge. Unfortunately, many of the 'plumber from hell' stories are true, but there are also thousands of qualified plumbers who are more than happy to do a fair day's work for a fair day's pay. How do you distinguish between the good, reliable plumber and the bad 'Mr Bodge'? The best option of course is to use a plumber who has been recommended by a friend. But if you can't do that, get some names out of the phone directory, phone up and ask about qualifications and guarantees. If a potential plumber gets angry or makes excuses, simply keep searching. A professional who is fully qualified will be only too happy to provide you with evidence of his credentials.

Quotes and estimates

Great care should be taken when getting a price from a plumber. There is often some confusion between a quote and an estimate. A quote is a fixed price. It is the same as going into a shop and asking about the price of a new piece of furniture. You will be given a price and this is the price you expect to pay. An estimate, on the other hand, is an educated guess at the likely cost. Beware the plumber who says something like, 'It's going to cost you somewhere between £2000 and £3000', because human nature being what it is, this will always mean £3000-plus. Always get a quote, not an estimate. It is fine if the plumber starts by giving you an estimate, but always insist on a written quote. This quote

10 Expert Points

THE FOLLOWING TEN POINTS WILL HELP YOU FIND A QUALIFIED PLUMBER.

1 RECOMMENDATION
The ideal option is to use a tried and tested plumber recommended by friends or neighbours. Some local authority planning departments will send you a list of qualified plumbers in your area, but they aren't allowed to make individual recommendations.

2 QUALIFICATIONS
Ask to see the plumber's certificate of qualification. If necessary, check their qualifications with the relevant qualifying authority.

3 PROFESSIONAL APPEARANCE
The plumber's transport and personal appearance can provide some clues. Although a clean van, displaying a trading name and address, and a crisp appearance aren't really any sort of guarantee of plumbing skills, they are an indication that the plumber is doing his best to create a good image and to keep your home clean.

4 TESTING SKILLS
A good starting point is to ask your chosen plumber to do a small but tricky task, such as repairing the main interior stopcock or perhaps fitting a cistern. If you are pleased with the quality of the work and the price, ask for a quote for a larger job.

5 QUOTES AND ESTIMATES
Give the plumber a list of items that need doing and ask for a quote. Beware the plumber who looks first at you and then at the sky, and gives you an 'estimate' on the back of an envelope. A good plumber will go away and work out a quotation. He will itemize and price the tasks and set out everything so that you have a clear understanding of what the job involves.

6 CONTACT DETAILS
An established plumber will have a contact address and a phone number. Beware the guy who will only give you his mobile phone number – it is just possible that he is a 'cowboy' with no fixed address.

7 GUARANTEES
A guarantee gives assurance, in black and white, that the plumber will put things right if they go wrong. However, if the plumber goes out of business you are stymied. That said, some plumbers are happy to back up their guarantees with insurance options.

8 SCHEMATIC PLANS
If the size of the task permits, the plumber should draw out a plan. This might not be much more than a pencil drawing, but it will set out the pipe runs, establish the position of valves and stopcocks, and so on. It will show you how the pipes are going to travel around the house, and indicate that the plumber is putting thought into the job.

9 REGISTRATION NUMBER
Most plumbers are members of an association or guild, such as the Institute of Plumbing, Association of Plumbers and Heating Contractors, or Council for Registered Gas Installers (CORGI). Ask for the individual's registration number and check it out either by phone or through the internet (www.corgi-gas-safety.com). By law, gas fitters must be registered with CORGI.

10 INSPECTION CERTIFICATES AND INSURANCE
The plumber should be happy to give you a signed certificate once the work is complete. Establish that this is the case at the beginning of the job. A certificated plumber should be insured. Ask to see evidence and make sure that it is valid.

should include VAT - but only if the plumber is VAT registered! If the plumber gives you a VAT number, check it out either at the local VAT office or by going to the website of Her Majesty's Customs and Excise (www.hmce.gov.uk). It is not unknown for cowboy plumbers to charge VAT even though they are unregistered. If you think that there is a problem about your plumber's VAT status, the website gives advice.

Once the work is complete, ask for a test certificate. File this certificate alongside the deeds of your property, so that when the time comes to sell your home, the certificate can be shown to prospective purchasers. In Scotland, and now more commonly in England, you have to produce a 'buyer's pack' for the intended purchaser, consisting of reports, surveys and certificates.

To recap, a quote is a fixed price given against your specification. The quote must be given in writing and should refer to 'the attached specification' (which should be stapled to the back of the quote). Take care when producing the specification. Carefully read the relevant sections of this book and make sure you have included everything you want. Go around the house with the plumber and look at the work before the quote is prepared. Ask the plumber to read through the specification. If he says that something is not such a good idea, wrong, or simply can't be done, and you think that his argument is reasonable, amend the specification accordingly

and send him a copy of the new specification to price. Working in this way, the plumber will know exactly what there is to do, and exactly how you expect the work to be finished.

Deposits

A well-established plumber shouldn't require any money prior to starting work; however, it is becoming increasingly common to be asked to pay a small amount of money as a deposit, especially if the plumber is starting out in business. The rest of the money should be paid as soon as the whole job is finished and you have the test certificate. Make sure that the details of these payment terms are written into the specification.

When to accept an estimate

Although a firm, fixed quote from the plumber is the ideal, many will only be willing to provide a very rough and ready verbal estimate. Their argument is that there are usually so many variables and unknowns (such as poor plaster, or bad plumbing hidden away in the wall cavities) that could push up costs, they really cannot be expected to give a fixed price.

So what do you do if you have made contact with three or more plumbers and they all veer away from giving a firm quotation? We could set up two ground rules: if it's a large job (more than a few hours' work), always insist on a quote; if it's a modest job (no more than a few hours' work), an estimate should be adequate.

Of course, it's not always the case that a plumber is going to cheat you, it might simply be that he hates paperwork and prefers a verbal agreement. Nevertheless, you always have to be on your guard. If you have no choice other than to go for an estimate, you must at the very least ask for a written agreement that clearly sets out the maximum price that you will have to pay for the work.

To recap, while the ideal is to obtain a fixed quotation in writing, if the task is only small and the plumber is only prepared to provide an estimate, it must be backed up with a written 'maximum price'. If the plumber will not agree to prepare such a document, it is best to keep looking for someone else.

Choosing from a shortlist of plumbers

Select three plumbers for your shortlist, give each your job specification, and ask for a written quotation. You will probably get three different prices – let's say, for example, £1000, £1500 and £2000. The difference between the lowest and the highest price is £1000. This gives you a guide price – the job ought to cost about £1500. Now ask yourself whether the lowest price is too low, or the highest price unreasonable. Beware, at this point, of always going for the lowest price. If possible, show the prices to friends and neighbours who have recently been through the 'plumbing experience', and take note of their comments. The difficulty comes when the prices are very close – say £1000, £1100 and £1200. In this instance, opt for the plumber who

seems to be on your wavelength. If you wish to include more plumbers on your shortlist, it might make the final choice easier.

Standard of work

The standard of the workmanship must ensure that the plumbing is safe and fit for its purpose. It must conform to certain regulations and codes of practice as set out by the various plumbing institutes and associations. Although there is no legal requirement for plumbers to belong to such a body, they should nevertheless provide you with a guarantee that states that the job is up to standard. Ask for this document just before you hand over the money.

What the eye can't see

As with most building jobs, such as bricklaying or electrical works, the problem with plumbing is that while you can see how some components are fitted (such as the taps and the pipes running around the skirting), there is no way that you can tell if concealed parts (such as the pipes under the floor) are up to standard. So if the components that you can see are not quite as they should be – perhaps a bit crooked – the likelihood is that the parts that are hidden from view will be totally unsatisfactory.

Do not wait until the job is done to air your concerns, but mention them the moment that you spot a problem. If you see that a tap is crooked or scratched, be bold and point it out. Don't let the job continue if you are unhappy about the standard of work.

19

That said, you mustn't hover or generally be a nuisance while the plumber is working. Remember that the work must be 'safe and fit for its purpose'. The best way of assessing the safety of a pipe, valve, tank etc. is to read through the relevant sections in this book and take note.

How to avoid cowboys

The problem with so-called 'cowboy' plumbers is that they are difficult to identify. They might look and sound like plumbers, and some might even think that they are plumbers, but the sad truth is that they are dangerous frauds who should be avoided.

There are two types of cowboy plumber. There is the honest but incompetent guy, and there is the ruthless guy who is out to take as much of your money as he can. They are both dangerous. The incompetent guy will perhaps give you a low quote that will inevitably result in a bad job, because the quote doesn't allow enough money to complete the work to a proper standard; alternatively he may have a higher opinion of his capabilities than is borne out by the job he turns in. The ruthless guy will knowingly do a bad job, take your money and swiftly move on. While we might feel sorry for one and dread the other, they are both dangerous.

Identifying the baddies

The following pointers might just help you identify and avoid cowboys.

- Never employ the chap who arrives at your door uninvited and claims to

be new to the area, doing a survey, has just done a job up the road, or is able to do a one-off job for a special low introductory price.

- Don't be overly impressed by smart clothing, big shiny vans and so on – you don't want to end up being the person paying for them!
- Avoid taking on the guy who wants large piles of cash up front.
- Stay away from the bloke who only has a mobile phone and won't provide further contact details.
- Try phoning a prospective plumbing firm during working hours (when you would expect the plumber to be out working for clients) to see if it is sufficiently well organized to be able to take and pass on messages.
- Be wary of the guy who is offering to do a swift job for cash.
- Avoid chaps who are only prepared to work out a rough and ready estimate on the back of an envelope.

Finance

The easiest and most convenient way to pay for a small job might be in cash, but the safest way to pay for large, expensive jobs is through a bank. Be very suspicious if your friendly plumber insists on cash. You should pay by cheque and get a written, dated and signed receipt.

There might be times when you need to sort out some finance for a large, expensive job, such as a complete refit of plumbing throughout the house. You have the choice of extending your mortgage and getting a loan that is secured on the property, or getting an unsecured loan that you pay back over 3-5 years, in much the same way as you might buy a car. Both options enable you to get the money at the lowest possible cost, and you will have the added benefit of knowing that cowboy plumbers will be much more wary of working for you if they know that a bank is involved.

Grants

Local authority grants are available to help with the cost of some plumbing projects, but they are not necessarily easy to come by. Elderly people, with little or no income and severe plumbing problems, are definitely eligible for help. The local authority has a duty of care to make sure that elderly and vulnerable people are safe in their homes. If you fall into this category, contact the council and tell them your age and situation.

To find out whether you qualify for a grant, make an initial phone call to your local planning authority and establish a contact name and address for future reference. Then send a letter to the appropriate head of department setting out what you consider is the problem. It is a good idea to send your letter by Recorded Delivery. Keep copies of any correspondence you send or receive. A council representative will call and arrange for a qualified plumber to look at the systems in your home. A written report will be sent to the authority. If the plumbing is in any way dangerous - for example, if lots of lead pipes are discovered - the work will be started immediately.

Disability grants

Disabled people are entitled to various grants for disabled facilities. If you need a toilet downstairs, for example, or a new heating system, there are local authority departments that are very happy to help. These grants are yours by right, so don't be intimidated by red tape, or tired voices at the other end of the phone. The secret of swift action is to get the name of a head of department. If the grant is for a priority case - the very old, very young, or housebound - there are special priority funds for eligible applicants. There are also emergency grants available in certain situations.

Who are you letting into your home?

Always ask for identification before you let strangers into your home. If you live alone, make sure that you have a friend with you for the initial call. Be wary about letting more than one person at a time into your house. Although this advice perhaps suggests that you will be battling against all manner of robbers, the majority of plumbers are just trying to do their honest best to make a living. If you are worried about finding someone trustworthy, the council may be able to provide you with a list of tried and trusted people to choose from.

Paying your plumber

If the job is small, costing no more than about £100, payment is straightforward. But when the bill runs into hundreds or thousands of pounds it is best to write an extra passage into the contract to cover how the money is to be paid. For example, you might pay one quarter of the total bill when the materials have been delivered to your address, one quarter when the pipes have been put in place, and the bulk of the money when the work is up and running and you have been given the certificate – say a few days after the work is completed. If the work fails in some way and the final lump of money remains to be paid, you have some leverage. State in writing the amount of money you are going to hold back, and your reasons for doing so.

Day-work payments
Occasionally a plumber might suggest that you pay him an hourly rate. This is a good option for a swift task such as mending a frozen pipe, but is not so good if the task involves open-ended jobs such as fitting a new bath. If there is a chance that a job could get complicated, avoid an hourly rate.

However, it may be worth considering in certain situations. Let's say that you want all your taps changed – a simple, straightforward task. The plumber might suggest that you provide the materials and he does the work (perhaps in the evenings) at an hourly rate. This is quite a good option on two counts. You will know precisely how you stand with regard to the cost of the materials, and the labour will cost no more than an agreed hourly rate. There won't be any hidden costs. Never give a plumber a blank cheque. Ask for receipts and then pay him.

Insurance

All responsible, qualified plumbers must be insured against injury, damage to your property and so on. This is particularly important for major jobs. Before work commences, make sure that the insurance details are set out on the contract.

As a homeowner you are to some extent legally responsible for the well-being of your guests, neighbours, passers-by and even the plumber. What would happen if the plumber fell off a ladder, damaged the carpets, injured one of your guests, or burned the house down? Phone your building insurance company to find out how you stand. If you have any concerns, take out extra insurance. Make sure that this extra insurance is in place for the whole duration of the project.

Health and safety

For the short time that the plumber is in your home, it will become a dangerous place, with ladders to trip over, holes to fall into, loft or cellar open, tools falling, and so on. Some of the materials and procedures are potentially dangerous too, involving heat, dust, gas burners, fumes and suchlike, so it is wise to steer clear of working areas. Keep away from open drains, open tins of chemicals and power tools. However, the plumber has a duty to leave your home in a safe condition at the end of each day. If, after a few hours' work, you see that there are burns on the skirting boards or whatever – with the plumber doing a lot of swearing – then be prepared for the inevitable problems.

Snagging

Snagging is the term given to the procedure of bringing a job to a satisfactory conclusion. On a small job this might involve no more that asking the plumber to clean some mess off a tap, but on a large job it might involve making dozens of small alterations.

Make a list of problems for the plumber and keep a copy for your records. For example, let's say that you have a deeply scratched tap, a weeping pipe and a chipped sink. You can either ask the plumber to make good as the work proceeds, or you can get him to bring the work to order just before the final payment is made. Let him know that you are keeping a list. Snagging doesn't involve you paying extra, it's just a way of ensuring the agreed job is completed to a satisfactory level. (You couldn't ask the plumber to fit six taps when you have both agreed on four. Or you could ask him to fit the extra two, but must expect to pay more for this.)

Trade agreements and disputes

Most average home plumbing jobs can be done for a relatively modest amount of money, but some jobs do run into many thousands of pounds. For major jobs like these, it's a good idea to make sure that your plumber belongs to the appropriate federations and quality-mark schemes. This will not help if the plumber is criminally negligent, and the warranties aren't worth much in financial terms, but they will promote the overall feeling that your chosen plumber is a

responsible chap who is going to do his best. Ask him if he belongs to such a body and check him out, just to make sure. If you think that your chosen plumber is in some way or another not quite up to the mark, then it might be a good idea to check out some fact or other in this book – say the best way to fit a new WC – and then gently ask him what he would do in such a situation. Generally make him aware that you know 'how many beans make five'.

DIY options

Of all the activities that do-it-yourselfers get up to, plumbing is one of the easiest and can save you the most money. Or, put another way, you could say that of all of the do-it-yourself activities, plumbing is uniquely easy in that just about every situation or task – from fitting a new tap or repairing a burst pipe, to fitting out a new bathroom or installing a shower, can be achieved using a kit. Just about every job can be done using the 'buying the bits and putting them together' approach. All the more strange, then, that plumbing is also an activity that most beginners veer away from. Turning off stopcocks, soldering pipes and all those other plumbing pleasures make an exciting and challenging weekend for some people, but for others the same tasks hold all the appeal of a nightmare.

So, should you have a go at DIY plumbing and save yourself a whole heap of money? The following section will help you decide whether to go it alone or employ a professional.

10 Expert Points

TO FIND OUT WHETHER YOU'RE SUITED
TO DIY PLUMBING, LOOK THROUGH
THIS TEN-POINT CHECKLIST.

1 TRIAL RUN
To a great extent, DIY plumbing
involves replacing existing components such
as taps or lengths of pipe. You can test out
your plumbing potential by having a trial
dry run – or perhaps we should say 'wet'
run! If you are keen but unsure of your
skills, start by replacing small items such
as washers. Then you could go on to
changing taps, fitting a valve in the toilet
system, mending gutters and so on.

2 TOOLS
You are going to need the right tools
for the job. Look through the tools glossary
on pages 162–167 before you get started.
While you can get all the tools from a
general DIY store, it's a better idea to get
them from a plumbing specialist, so that
you can benefit from their expertise.

3 ADVANCE PLANNING
Always plan a job in advance. If you
have decided to change a tap, have you
bought a new tap of the right size? Do you
possess the correct tools to carry out the
task? Have you allowed yourself enough
time to do the work? For how long can you
keep the household without water? Does the
stopcock work? Write up an 'order of work'
plan that anticipates all eventualities.

4 IF THINGS GO WRONG
Always prepare an escape plan, just
in case things go wrong. If you have chosen
to fit new washers late on Saturday night,
what are you going to do if the washers are
the wrong size or pattern? Are you going to
run around the house shouting, have you
purchased alternatives, or have you already
checked that a supplier will be open on
Sunday morning, just in case?

**5 MAKE THINGS AS
STRESS-FREE AS POSSIBLE**
Plan work so that there will be the minimum
of stress. Allow your children to watch if
they want to, as long as they keep quiet.
Don't start a project late at night. Find out
whether the plumbers' merchant is open
the next day. Make sure that you have an
adequate supply of drinking water, just in
case you cannot turn the water back on
when you've finished working.

6 ORDER OF WORK
Do as much as you can before turning
off the water. For example, when replacing
a length of existing pipe complete with a
valve and bends, make up most of the pipe
before you turn off the water.

7 PRACTISING TECHNIQUES
If you are keen but nervous, practise
various techniques such as bending a pipe,
fitting a compression joint, soldering and
so on. This will build up your confidence
before you start plumbing for real.

8 TURNING THE WATER OFF
Many plumbing projects require you
to turn off the water and drain the system.
However, this is the point at which many
raw beginners panic and decide to opt out.
It's a good idea to practise this sequence
so that you can do it with confidence when
the real thing comes along.

9 ORDER, NOT CHAOS
Successful plumbing is achieved by
crisp, clean procedures: clean pipes, clean
joints, and the correct, properly-functioning
tools close at hand. Are you the sort of
person who likes order and procedures?

10 HELPERS
It's always a good idea to have a
helper. Do you have a friend to help and
support you along the way?

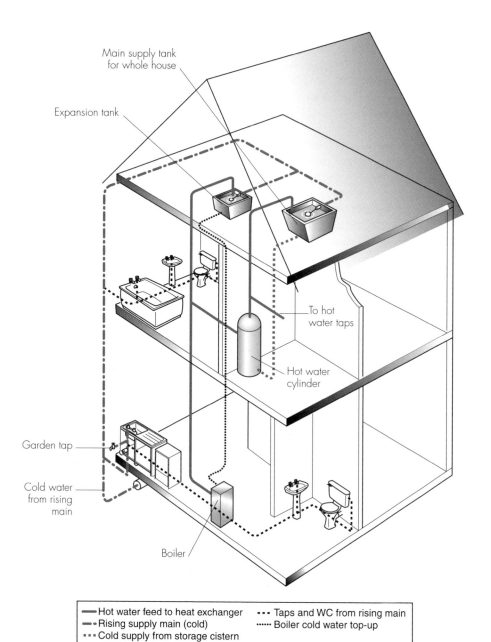

Main supply tank for whole house

Expansion tank

To hot water taps

Hot water cylinder

Garden tap

Cold water from rising main

Boiler

——— Hot water feed to heat exchanger	• • • Taps and WC from rising main
—•—• Rising supply main (cold)	•••••• Boiler cold water top-up
• • • Cold supply from storage cistern	

A direct cold water system with the cold water
to all taps and fittings coming directly off the rising main.

CHAPTER 3

Systems

Household plumbing systems

Most modern homes have four plumbing systems.

- A direct or indirect cold-water system that brings cold water in and distributes it through the system.
- A hot-water system that heats water and then distributes it to the various hot-water taps in the system.
- A hot-water central-heating system that heats the house.
- A waste system that collects and disposes of sewage, wastewater and rainwater. In country areas that are away from public sewers, this system will include a cesspit or septic tank for household waste, and individual soakaways for rainwater.

The direct cold-water system

Mains water comes into your home under pressure through a pipe called the rising main. In the direct system, this instantly branches off into a dedicated system of pipes that directly feed the cold-water storage tank in the loft, cold-water taps and outlets around the building. All the outlets are at mains pressure. (Some modern direct systems take all water from the mains and do not feature a storage tank.) This system has, to a great extent, been superseded by the indirect cold-water system.

If you intend to repair, modify or extend your plumbing, it is important to know whether or not you have a direct cold-water system. You should also be aware of the advantages and disadvantages of the system.

The indirect cold-water system

With the indirect cold-water system, the mains water comes into your home under pressure, and then the rising main pipe splits into two. One pipe serves the cold tap in the kitchen, and the other pipe feeds the cold-water storage tank in the loft. All other systems in the house are fed (at lower pressure, and by gravity) from the cold-water storage tank. Drinking water should only be drawn from the cold tap in the kitchen.

If you intend to repair, modify or extend your plumbing, it is important to know whether or not you have an indirect cold-water system. You also need to be aware of the advantages and disadvantages of the system.

The hot-water system

In most houses, the hot-water system is made up of three units: the water heater (a boiler powered by gas, solid fuel or oil), the hot-water cylinder and the cold-water storage tank in the loft.

Direct hot water

Traditionally, hot water was heated directly – water from the cold-water storage tank in the loft fed down to the bottom of the hot-water cylinder and then on down to the boiler, where it was heated. When the water got hot, it rose from the boiler up a pipe to the top of the hot-water cylinder, from where it was used to

10 Expert Points

TO FIND OUT WHETHER YOU HAVE A DIRECT COLD-WATER SYSTEM, LOOK THROUGH THIS TEN-POINT CHECKLIST. IT ALSO DESCRIBES THE SYSTEM'S ADVANTAGES AND DISADVANTAGES.

1 AGE OF THE PROPERTY
For the most part, only older properties have a direct cold-water system. Was your home built before the 1950s? If so, it probably does.

2 PIPES FEED OUTLETS DIRECTLY
A direct system can be recognized by the fact that the pipe from the main interior stopcock branches off into pipes that directly feed the water storage tank in the loft and the various cold-water taps, outlets and toilets around the house.

3 TOILET TEST
To test whether you have a direct system, turn off the main stopcock and flush the various toilets. Do the WC cisterns fill up? If they do not, you have a direct system.

4 TAP TEST
If, when the water supply company turns the water off in the road, your cold-water taps stop working, you have a direct system that takes water for the taps and storage tanks directly off the rising main.

5 SOUND TEST
When you run water from a cold tap in the bathroom, can you hear hammering or hissing? If so, the likelihood is that you have a direct system.

6 CONDENSATION TEST
Where the cold-water pipes run through cold spots in the house (for example, under the kitchen sink, or perhaps in a cold area behind a WC cistern), do the pipes drip with condensation? If so, you probably have a direct cold-water system.

7 FEED PIPE TO HOT-WATER CYLINDER TEST
If there is just a single cold-water feed pipe running from the bottom of the cold-water storage tank in the loft to the hot-water cylinder, you have a direct system.

8 ADVANTAGE OF A DIRECT SYSTEM
Any cold-water tap in the house will provide water that is safe for drinking. (In an indirect system, only water from the kitchen tap should be drunk.)

9 DISADVANTAGE OF A DIRECT SYSTEM
A distinct disadvantage of the direct system is that there is a slight chance that contaminated water from the toilet could siphon back into the drinking water. However, a modern direct system with non-return check valves will prevent drinking water being contaminated.

10 REPLACEMENT OF A DIRECT SYSTEM
If, after making all the tests described, you find that you have a direct system, the likelihood is that it is old and needs replacing. This needs thinking about. Ask the water supply company for advice.

feed the hot-water taps. If the hot water in the top of the tank wasn't drawn off, it cooled down, then fell to the bottom of the cylinder, and then on down through a pipe and back to the boiler to be heated again.

Indirect hot water
Over the years, the direct hot-water system has gradually evolved into a system where the water is heated indirectly, and used to feed both the hot-water taps and radiators.

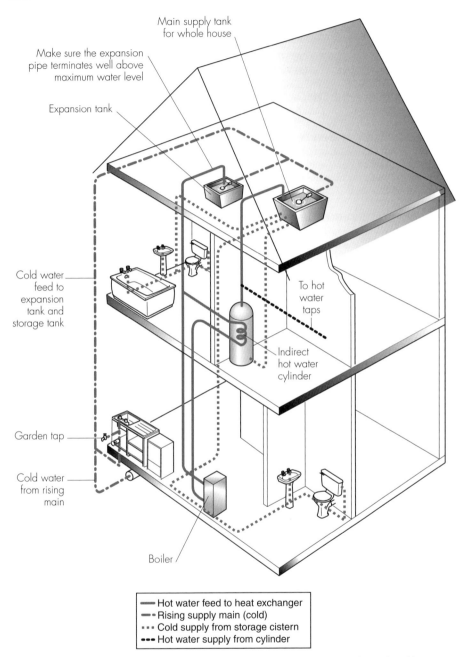

- Hot water feed to heat exchanger
- Rising supply main (cold)
- Cold supply from storage cistern
- Hot water supply from cylinder

A typical indirect hot and cold water system with bathroom and toilet fed with cold water from the storage tank (for clarity, hot water supply pipes to taps are not shown).

10 Expert Points

TO FIND OUT WHETHER YOU HAVE AN INDIRECT COLD-WATER SYSTEM, LOOK THROUGH THIS TEN-POINT CHECKLIST. IT ALSO DESCRIBES THE SYSTEM'S ADVANTAGES AND DISADVANTAGES.

1 AGE OF THE PROPERTY
Most modern properties have an indirect cold-water system. If your house was built some time during the last 50 years, it probably has such a system.

2 TWO MAIN PIPES
An indirect system can be recognized by the fact that the pipe from the main interior stopcock goes directly up to a cold-water storage tank in the loft, and there is a single secondary pipe branching off to feed the kitchen cold-water tap.

3 TOILET TEST
To test for an indirect system, turn off the main stopcock and flush the various toilets. Do the WC cisterns fill up? If they do, you have an indirect system.

4 TAP TEST
If, when the water supply company turns the water off in the road, your bathroom cold-water taps continue to work, you have an indirect cold-water system.

5 SOUND TEST
When you run water from a cold tap in the bathroom, can you hear hammering or hissing? If not, the likelihood is that you have an indirect system.

6 CONDENSATION TEST
Where the cold-water pipes run through cold spots in the house (for example cold areas in the bathroom), do the pipes drip with condensation? If not, you probably have an indirect system.

7 FEED PIPES FROM THE COLD-WATER STORAGE TANK
If there is more than one cold-water feed pipe running from the bottom of the cold-water storage tank in the loft, you have an indirect cold-water system.

8 ADVANTAGE OF AN INDIRECT SYSTEM WHEN WATER IS CUT OFF
One advantage of an indirect system is that if for some reason the water needs to be cut off, there will be a good supply of water in the cold-water storage tank to feed the cold-water taps and WC flush.

9 ADVANTAGE OF AN INDIRECT SYSTEM: NO CONTAMINATION
An advantage of the indirect cold-water system is that there is no chance that contaminated water from the toilet will siphon back into the drinking water.

10 MODERN PLUMBING
If you find that you do have an indirect system, the good news is that your plumbing is relatively modern and will not need to be replaced. If you are in any doubt about the type of system that has been installed in your house, then call in a plumber to carry out a proper inspection.

In the indirect system, there is *usually* another water tank in the loft, as well as the main cold-water storage tank that supplies water to the cold taps and feeds the outer part of the hot-water cylinder: a smaller feed-and-expansion tank dedicated to the heat exchanger in the hot-water cylinder. Each tank has its own pipe run. A pipe runs up from the boiler, through the heat exchanger (a coil) in the centre of the hot-water cylinder, and then on to feed the radiators. Another pipe starts at the feed-and-expansion tank, and

10 Expert Points

THE FOLLOWING RECOMMENDATIONS
WILL HELP YOU TO UNDERSTAND YOUR
PLUMBING SYSTEM AND TO FIND OUT
WHETHER YOU HAVE A DIRECT OR
INDIRECT HOT-WATER SYSTEM.

1 CHECKING PIPES
If there is a maze of pipes in the airing
cupboard (where the hot-water cylinder is),
it's a good idea to feel if they are hot
or cold and mark them with blue or red
insulation tape accordingly. This will give
you some idea of their function.

2 IDENTIFYING OLD PIPEWORK
In an old house, you may find old
pipework has been left in place in the loft
and in various cupboards around the house.
This can be confusing. If you do see
pipework that is obviously unused and open-
ended, it's a good idea to mark it as such,
or better still to remove it.

3 OLD HOUSES
If you live in an old house with a basic
hot-water system, with perhaps a back-boiler
heated by an open fire, the likelihood is that
you have a direct system.

4 AGAS AND RAYBURNS
If you live in a cottage and have a
very basic hot-water system (for example an
old Aga or Rayburn) with hot-water taps in
the kitchen and bathroom, and you rely on
the Aga/Rayburn rather than radiators to
heat the house, then the likelihood is that
you have a direct system.

5 PIPES IN THE HOT-WATER CYLINDER
Follow the hot- and cold-water pipes
up from the boiler to the hot-water cylinder.
Do they run directly into the cylinder without
branching? If so, you have a direct system.

6 CONFUSING LABELS
You cannot always believe existing
labels lurking on your plumbing system.
We have seen a cylinder labelled 'indirect',
connected to a direct system. Don't take
anything for granted.

7 MODERN HOUSES
If you live in a modern house with a
large boiler that heats both the hot taps
and the central heating radiators, you
almost certainly have an indirect system.

8 LOOKING IN THE LOFT
Go up into your loft and look around.
Do you have one storage tank or two? If
there are two, each with two pipes, you
have an indirect system.

9 BROKEN HEAT EXCHANGER
We have seen indirect systems where
the heat exchanger inside the cylinder has
broken, resulting in the system becoming a
direct one. Be mindful of this possibility.

10 MAKING SURE
To be absolutely certain whether your
system is direct or indirect, run through all
the tests suggested here. If you are in any
doubt, call in a plumber to inspect the
system and advise you.

expansion tank, and goes down to
feed the outer part of the cylinder and
on to the hot-water taps. With the
indirect system, the hot water in the
heat exchanger coil within the
cylinder heats the water in the outer
part of the cylinder, with the two
bodies of water never meeting. Both
pipe runs are protected from
overheating by an expansion pipe that
runs to the feed-and-expansion tank.

Modifications

If you plan to repair, modify or extend

your plumbing, it is important to know whether you have a direct or indirect hot-water system. You also need to be aware of the advantages and disadvantages of both systems.

Heating systems

Although several different heating systems are commonly used, we will only be discussing 'wet' systems, which have water-filled radiators. In wet systems water is heated in a boiler and pumped to each radiator around the house in small-bore pipes. Additives may be put into the water to inhibit rusting in the radiators. The pipework is usually copper. Most is 15 mm in diameter, although some can be 'microbore', which is about 10 mm. Microbore is not as efficient, but is much easier to fit. Older houses sometimes have a 'gravity' system, which does not have a pump and the hot water rises naturally and circulates by itself. The water in the system is much cooler when it returns to the boiler, where it is reheated and then circulates through the system again.

Boilers

The boiler is usually heated by mains gas or oil (kerosene). Oil is delivered by tanker and stored in a plastic or steel tank outside the house, either as an individual tank sited in the garden, or in a purpose-built block serving a small community. Boilers can also be heated using electricity, coal or propane gas. Propane gas is stored either in a large cylindrical tank in the garden or purchased in small red cylinders (available at petrol stations).

Choosing a fuel

Your eventual choice of fuel for heating will depend upon four factors: the availability of the different types of fuel, whether or not your house has a chimney, whether or not there is someone at home during the day, and last but not least, your physical strength and willingness to carry fuel into the house from outside.

Is there mains gas in your road? Do you live in an isolated community away from a source of solid fuel? Are you allowed to burn solid fuel in your area? Is there a supplier delivering oil in your area? Is there room in your garden for an oil tank or solid-fuel bunker or shed? Is there someone at home during the day to burn logs and coal to fuel the boiler (other options might be more efficient if the house is usually empty during the day)? Are you willing to carry heavy solid fuel or gas cylinders around, especially in winter? You might have a chimney and relish the idea of an open fire, but chopping logs or heaving coal into the house in the middle of winter might not be such an appealing idea.

Once you have assessed all these factors, the real options become quite obvious. If your house doesn't have a chimney, you cannot use solid fuel, but you could have a gas- or oil-fired boiler fitted with a balanced flue vent. If mains gas is not available in your area, but your house does have a chimney, you are limited to using either oil or solid fuel, even if your garden is too small to easily accommodate a storage tank or fuel shed or bunker for storing coal, logs or other solid fuel.

10 Expert Points

THE FOLLOWING POINTS WILL HELP YOU TO TAKE THE RELEVANT FACTORS INTO ACCOUNT BEFORE PLANNING A HEATING SYSTEM (OR INHERITING ONE IN A NEW HOUSE).

1 SYSTEM COSTS
Which is the cheapest system to install? As a rule, it is the one in most common use. At the time of writing, this is mains gas central heating with 15 mm pipes and steel panel radiators. (However, many experts think that small electric convector heaters are a good option. Even though they are expensive to run, a small system can be installed at very low cost.)

2 RUNNING COSTS
Which system will be the cheapest to run? At the moment, mains gas is the cheapest heating fuel, followed by oil, propane, coal and electricity, in that order.

3 GAS CENTRAL HEATING
If you have a gas fire, you can have gas central heating. The gas pipe to your house will have a bore of at least 24 mm. This will supply enough gas to permit central heating, cooking and open gas fires all to be in operation at the same time.

4 DIY INSTALLATION
The design of the system should be done by an expert, but the fitting of it can be tackled by a DIY enthusiast. However, you must not tamper with the gas supply in any way. A CORGI engineer must connect the gas to the boiler and test it.

5 CORGI ENGINEERS
There are many bogus gas fitters around, and some even have fake ID cards, so it is best to check that the potential engineer is registered by CORGI (the Council for Registered Gas Installers). Ask for the engineer's registration number and address before he comes to your house.

Use the internet to check this information on the CORGI website (www.corgi-gas-safety.com), which will tell you if the engineer is approved. Ask for an ID when the engineer comes to do the job.

6 MOVING HOUSE
When buying a house, ask for the heating system to be demonstrated. You will have no come-back once you have moved in, and most surveyors do not include checking the operation of the heating system in their structural survey – so beware!

7 SITING AN EXTERIOR TANK
If you are fitting an oil or gas tank outside, call the fuel supplier and ask how long the feed pipe on the lorry is. Don't put the tank too far away!

8 PLANNING PERMISSION
Check whether you need planning permission to fit a fuel tank. Your house may be listed and special conditions may apply.

9 NEW SYSTEM DETAILS
If you are having a system fitted, make sure you have all the details in writing from the heating engineer, including the make of the boiler, the heat output measured in BTUs (British Thermal Units), the number of radiators together with their dimensions and positions, the size of the pipework, whether the pipes are on the surface or hidden, whether the decorating/flooring etc. is to be made good, and the predicted temperature for each room.

10 FULL WORKING ORDER
Make sure that the installer's quote is for a system that is 'fully installed, connected, tested, commissioned and in full working order'. Ask for the operator's manual for the boiler. Many installers are happy to wait until you have tested out the system before you pay in full. Try to build a 'pay in full' clause into the contract.

The waste drainage system

The waste drainage system is used to dispose of sewage from the toilets and wastewater from the baths, sinks, washbasins and washing machines. This is directed, via U-bend traps, to various collection points. From there the waste is sent down a large-diameter, vertical pipe (sometimes built inside the house) and on through an underground pipe and inspection chambers (where it is joined by rainwater), before being fed into a public sewage system. The sewage is flushed along its way by both the wastewater from the house and the rainwater from its roof.

In country areas, where houses are remote from public sewers, the waste system will include either a cesspit or septic tank to cope with the waste from the house, and individual soakaways (one from each of the rainwater downpipes) to take rainwater. If you live in the country or enjoy gardening, you could install rain butts instead of soakaways.

It is vital to understand how the waste drainage system in your house works, especially if you plan to build an extra bathroom, or if you live in the country, when it's essential that you make extra plans to accommodate the wastewater that you will be generating. You won't have a real problem if you live in a town, since you can simply run all the waste into a sewer, but if you live in the country you will have either a septic tank or, if you are less fortunate, a cesspit. The problem is, of course, that modern appliances such as washing machines and dishwashers send out vast quantities of wastewater, and this is not at all suitable for a cesspit or septic tank. Of course, if you have lived in the same country house for many years you will have a good understanding of just how your wastewater system works, but if you have just moved into an old house or cottage and have never encountered either a septic tank or a cesspit, then the best way forward is to have a poke around and to try to identify just what it is that you have lurking at the end of the garden. Are there any unpleasant smells? Is the ground soft and wet? Talk to the neighbours and generally see if you can build up a picture of what is going on. Of course, if the system hasn't been touched for years, then the likelihood is that it is a septic tank, but if it has been emptied three or four times in the past year, then the outlook isn't so good as it is most probably a cesspit. The Expert Points overleaf will provide most of the answers to your questions.

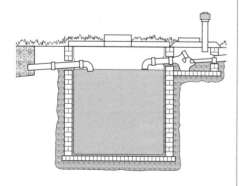

A traditional cesspit system
(see also page 65).

33

10 Expert Points

THE FOLLOWING POINTS WILL HELP YOU UNDERSTAND HOW THE WASTE DRAINAGE SYSTEM WORKS.

1 TYPE OF SYSTEM
The drainage system you have will depend on whether you live in a house or flat, whether it is old or new, and whether it is situated in a town or the country. Systems differ slightly in order to suit local conditions and to conform to various acts and laws. So, you might have a two-pipe system, a single-stack system, a communal system, a cesspit system or a septic tank system.

2 IDENTIFYING A TWO-PIPE SYSTEM
If you have an old house with a two-pipe system, waste from the toilets will feed into a large-diameter soil pipe, and down to an underground inspection chamber. Wastewater from baths, sinks and washbasins flows (via separate pipes) into a hopper and then a trapped gully before reaching the inspection chamber.

3 IDENTIFYING A SINGLE-STACK SYSTEM
In the single-stack system, the waste from the toilets, baths, washbasins and kitchen sink all comes down the same large-diameter pipe and is channelled into the underground inspection chamber. (Sometimes the water from the kitchen sink discharges into its own trapped gully.) In older houses, this stack pipe will be on the outside of the building, but in modern houses it might be hidden from view inside the house.

4 IDENTIFYING A COMMUNAL SYSTEM
In a block of pre-war flats, which might have either a two-pipe or a single-stack system, the system is made more complicated by the fact that the waste shares a communal pipe run. The residents will have to share the costs and responsibilities of the system.

5 COMMUNAL ETIQUETTE
If you have a communal system, you mustn't make changes or even undertake any DIY tasks without talking to other flat owners. If you have any doubts, look at your contract or lease before embarking on any alterations.

6 IDENTIFYING A CESSPIT SYSTEM
A cesspit is no more than a large brick- or concrete-lined hole that acts as a holding chamber for waste. Usually, there will be an inspection chamber before the pit, and a ventilator on the top of the pit. In action, the waste goes through the inspection chamber and into the cesspit. A cesspit and a septic tank are not the same thing. Be wary of cesspits that have been extended to incorporate an overflow.

7 DISADVANTAGES OF A CESSPIT
The trouble with a cesspit is that it is smelly and constantly needs emptying. This can be a huge problem in areas where the top soil is heavy clay.

8 IDENTIFYING A SEPTIC TANK SYSTEM
A septic tank system has an inspection chamber, a large tank or chamber, and a filtration system. The waste goes through the inspection chamber and into the large tank, where the solids fall to the bottom and the water flows off through the filtration tank and on into the soil.

9 ADVANTAGE OF A SEPTIC TANK
A good septic tank hardly ever needs emptying. If it doesn't smell and the topsoil isn't oozy, leave it alone.

10 SEPTIC TANK MAINTENANCE
If a septic tank smells, have it emptied and wash or replace the layers of sand and gravel in the filter bed.

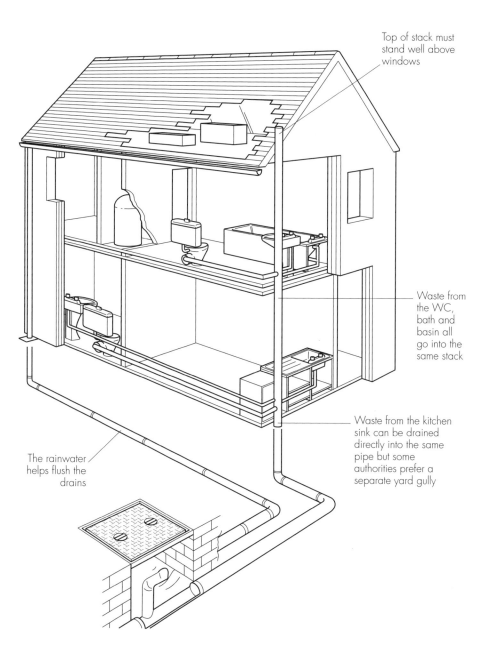

Top of stack must stand well above windows

Waste from the WC, bath and basin all go into the same stack

Waste from the kitchen sink can be drained directly into the same pipe but some authorities prefer a separate yard gully

The rainwater helps flush the drains

Drainage layout of a typical system.

CHAPTER 4

Hardware

Materials and fixtures

All the pipes, clips, joints, tanks, cisterns, taps and so on that make up the plumbing system in your home are known as hardware. With plumbing items, just as with everything else, you get what you pay for. Always buy the best-quality materials and fixtures. If you are going to all the trouble and expense of installing new plumbing, whether you are doing it yourself or bringing in a plumber, it makes sense to use high-grade hardware to ensure that the system will last at least 20 years. Of course, this doesn't mean that you need gold-plated bath taps, or a bath tub that is big enough to take the whole family, just that you should get good, straightforward fixtures and fittings that conform to all the British Standards.

Getting a good deal

You can cut costs by buying all the pipes, joints and fittings in bulk. Work out the number of taps, joints and metres of pipe required. Remember that the items must be compatible, and all manufactured to the same British Standard. Then phone around various suppliers for prices. Ask for their very best price on the whole package, and then on groups of items within the package. Bear in mind that some specialist plumbing suppliers only sell items of a certain type, such as plastic pipes and fittings, or copper pipes and fittings, and so on.

It might be that you can get the best deal by getting all the copper from one supplier, and all the large items such as baths and washbasins from another supplier, and so forth. However, a traditional, long-established plumbers' merchant might be the best

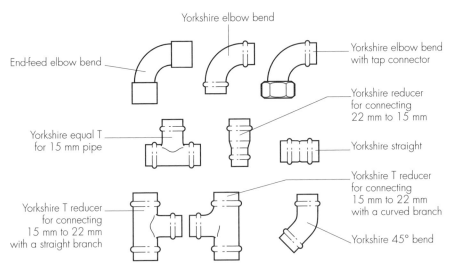

Yorkshire elbow bend

End-feed elbow bend

Yorkshire elbow bend with tap connector

Yorkshire reducer for connecting 22 mm to 15 mm

Yorkshire equal T for 15 mm pipe

Yorkshire straight

Yorkshire T reducer for connecting 15 mm to 22 mm with a straight branch

Yorkshire T reducer for connecting 15 mm to 22 mm with a curved branch

Yorkshire 45° bend

Capillary joints.

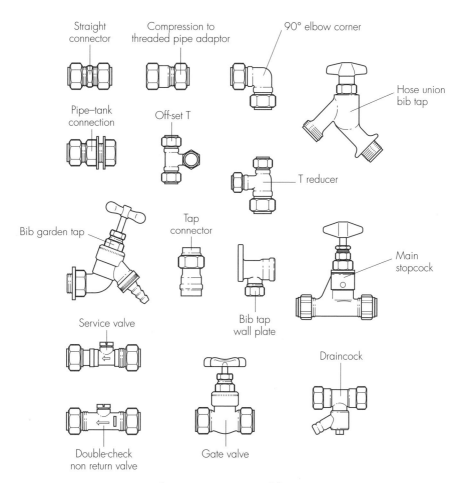

Straight connector
Compression to threaded pipe adaptor
90° elbow corner
Hose union bib tap
Pipe–tank connection
Off-set T
T reducer
Bib garden tap
Tap connector
Main stopcock
Service valve
Bib tap wall plate
Draincock
Double-check non return valve
Gate valve

Metal compression joints and fittings.

solution if you are a beginner. They may be a little more expensive, but are generally happy to give advice and answer your queries.

Pipes and joints

It is not so long ago that the plumbing systems in most houses were fairly straightforward. Ambitions did not extend further than having hot and cold water and an inside toilet, and the system was made up of lead pipes, brass fittings, ceramic sinks and cast-iron baths. Now, however, there is such a huge variety of hardware available, and so many factors to consider when making changes or additions to a plumbing system, that the process of choosing materials can be extremely confusing.

37

10 Expert Points

THE FOLLOWING POINTS WILL HELP TO GUIDE YOU THROUGH THE PIPE AND JOINT MAZE, AND ASSIST IN IDENTIFICATION AND SELECTION.

1 LEAD PIPES
Lead is potentially poisonous. If you have lead pipes, they should be removed.

2 GALVANIZED STEEL PIPES
Galvanized steel, now considered to be an outdated product, was traditionally used for underground work. To a great extent, it has now been replaced by plastic. If you have galvanized pipes, they are likely to be near the end of their life.

3 CAST-IRON PIPES
Cast iron is now mainly used in restoration work for gutters and stack pipes. It is both expensive and difficult to handle.

4 STAINLESS-STEEL PIPES
Stainless steel was once only used for public toilets and facilities, but stainless-steel pipes are now increasingly being used as a design feature alongside stainless-steel fixtures such as sinks and baths. This material, although modern and trendy, is rather difficult to work.

5 BRASS AND GUNMETAL ITEMS
Brass and gunmetal are used for joints and taps rather than for pipes.

6 COPPER PIPES
Copper is widely available, hardwearing, easy to bend, easy to solder, and generally good for both cold- and hot-water systems. Most plumbers would say that copper would be their first choice.

7 COMPRESSION JOINTS
Compression joints come in all manner of sizes, angles, junctions and forms to fit various pipe types, sizes and layouts. When you are making a joint in a difficult or awkward situation (such as in a loft, or where an existing pipe is still partially full of water), compression joints are the option to go for.

8 CAPILLARY JOINTS
Copper capillary joints come in a whole range of sizes, angles, junctions and forms to fit your chosen size and layout of copper pipes. There are two types – the 'Yorkshire' joint that has a ring of solder built into its structure, and the 'end-feed' joint that requires you to feed solder into the joint. Capillary joints are inexpensive, tidy and easy to use, but can only be used when the pipes being jointed are completely dry.

9 COMBINATION JOINTS
Combination joints, made in a mixture of brass and copper, are used in situations where you need to create a link by means of a joint that has a compression or screw-fix fitting on one end and a soldered fitting on the other. For example, at the point where a long length of solder-jointed pipe comes to meet a tap, you would need to make the link with a joint connector that has a capillary joint at one end to fix it to the pipe, and a cap-nut screw-fix joint at the other end to fix it to the tap.

10 FITTINGS
To complete the plumbing system, fittings are required. Most fittings are cast in brass, but those that are on show, such as taps, tend to be chrome-plated.

Cutting copper pipe

Copper pipe is widely available and generally good for both cold- and hot-water systems. You can cut it using either a fine-bladed hacksaw with more than 120 teeth per inch (sometimes called a 'junior' hacksaw), or a proprietary rotary pipe cutter.

It is important to cut the ends of a pipe so that they are exactly square and will fit neatly into the coupling. A sloping cut on the end of a pipe means that you might not be able to solder it properly, or, if fitting a compression joint, you might not get a good seal. In order to achieve a perfectly square cut every time, a rotary pipe cutter (or tube cutter) is the best option. You will find, however, that a rotary pipe cutter is only suitable if you can get all the way around the pipe. If the pipe is tight against a wall, you will have to use a fine-bladed hacksaw.

Cutting a pipe with a hacksaw

The most important thing about cutting a pipe is to get the length

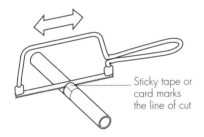

Cutting pipe with a fine-toothed 'junior' hacksaw.

exactly right. There is an old saying in the trades (woodwork, joinery, electrics and, of course, plumbing) that goes something like, 'Measure twice, cut once'. Copper pipe is expensive and once you've cut it to the wrong length, your only option is to throw the pipe away and start again or joint it with a straight connector.

Measure the pipe and wrap a small piece of paper tightly around it, aligning the edge of the paper with the line of cut. Place the pipe across a couple of strong kitchen chairs or a workbench – so that it is well supported and not 'whipping around' – and use a fine-bladed hacksaw to cut it, following the line of the edge of the paper. Do not put too much pressure on the hacksaw as it will bite into the copper and twist the blade. Use blades that have more than 100 teeth per inch (this will be marked on the blade packet as '100 TPI') or equivalent. Once the pipe has been cut, take off the burr (frayed edge) with a round file on the inside and a flat file on the outside. The files should be the smoothest that you can buy.

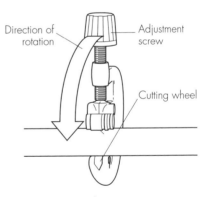

Cutting pipe with a rotary pipe cutter.

Cutting a pipe with a rotary pipe cutter

Although this method is beautifully simple, you must still take care with measuring and marking the pipe as previously described. Make sure you have allowed enough length at either end to fully enter the joint.

There are two types of rotary pipe cutter. One looks rather like a G-clamp, with two small rollers and a cutting wheel, while the other looks like a round, plastic doughnut or ring with the wheels and cutter blade on the inside edge. The ring type is the best, but each cutter is only good for one pipe size, so for most plumbing jobs you'll need to buy both a 15 mm and a 22 mm model.

Both types of cutter work in the same way. They are placed around the pipe so that the cutting wheel lies along the line you have marked. The cutter is then rolled around the pipe and gently screwed tighter until the pipe separates. The ring-type cutter is squeezed by hand as it is rotated around the pipe.

After cutting, take out the burr on the inside of the pipe using either a round file or the triangular reamer that is to be found on the other end of the clamp-type pipe cutter.

Forming bends in copper pipes

There are several methods of forming a bend in copper pipework. The two most common are a long-radius bend, where the pipe itself is bent, and making a connection between two straight pipes with a soldered joint or a compression joint.

Long-radius bends

The long-radius bend is made by physically bending the length of copper pipe, a method favoured by plumbers as it avoids the cost of purchasing a pre-made bend. A

bending spring or a bending machine is used, which prevents the pipe kinking like a broken drinking straw.

Bending springs
Different sizes of bending spring are available to match standard pipe diameters. The spring fits into the copper pipe and stops it kinking. Bending springs are available from DIY stores, but can also be hired from tool-hire centres. Once protected by the spring, the pipe can be simply bent over your knee.

The maximum diameter of pipe that can be bent with a spring is 22 mm. To bend larger pipes, a bending machine must be used. These are more expensive to buy, but they are also available to hire.

(If you are constructing quite a short section of pipework, it might be somewhat cheaper to buy end-feed, solder-on bends. These will give the pipe a tidy, sharp bend that will fit tightly into a corner.)

Using a bending spring
Method
1 Mark the pipe carefully, allowing an extra few centimetres for trimming off. If the required bend is not near the end of the pipe, you will need to attach string to each end of the bending spring so that you can position it correctly. The string should be about the size of thick bootlaces and strong enough to pull hard without it breaking. The reason for this is that, although the spring will slide in when the tube is straight, it will not come

Grip the end of pipe
and bend it round a post
or over your knee

Bending pipe with a spring.

out so easily because the bent pipe will grip it tight. Sometimes it is better to grease the spring with a little pure petroleum jelly (do not use motor-lubricating grease, as this could contaminate your drinking water).

2 Lay the string and spring beside the new pipe, making sure the spring lines up with the area to be bent. Tag the string or spring with masking tape to show the position of the two ends of the pipe, so that the spring is in the correct position when inserted. Attach a large nail to one end of the string, hold the pipe upright and feed the nail and string in at the top of the pipe – gravity will do the rest!

3 Bend the pipe over your knee or under your foot, as though snapping a stick. (Soften pipes that are made of thick or hard metal by playing a blowtorch over the area to be bent until it is red-hot. Let the pipe cool very slowly at air temperature to anneal or soften it. This will not harm the pipe.)

4 The spring may be difficult to remove from the pipe, even by pulling hard. If this is the case, put a nail or screwdriver through the end of the spring, if it is accessible, and wind it clockwise. The spring will coil more tightly and become smaller, and can then be pulled out. Wear a heavy glove for this just in case you slip and the screwdriver or nail spins back.

Using a bending machine
Method

1 Mark the pipe with a permanent marker to show where the start and end of the bend should be. The bending machine will have a set of forms – usually 15 mm, 22 mm and 28 mm – for each standard diameter of pipe. Each set of forms will have a straight and quadrant (or curved) part.

2 To make a 90° bend, place the pipe in the quadrant part so that it hooks under the machine. Place the straight part on the other side of the pipe and push the handle of the machine around. The straight part will bend the pipe into the curve of the quadrant.

3 Angles other than 90° can be made by pushing the handle of the machine only part of the way around. On some machines, various choices of angle are marked on the quadrant. You will find that most pipes will have to be 'over-bent' slightly as they spring back when taken out of the machine. After you have made the

41

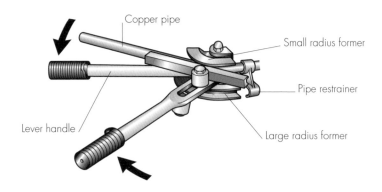

Bending pipe using a machine.

first bend or two, you will know how much to over-bend the pipe to achieve the desired result.

4 It is not possible to make a bend right at the very end of a length of pipe, because the machine will crush the end of the pipe into a slightly oval shape, and consequently make the fitting of joints very difficult. You need at least 25 mm of straight pipe on either side of the bend in order to connect fittings or couplings.

Soldering joints and connectors in copper pipe

By far the easiest method of fitting a bend (or elbow joint), T-joint or straight connector is to make a capillary joint where solder is used to fill the tiny space between the pipe and joint sleeve. Soldered joints are small, neat, tidy and cheap; the downside is that you will need to use a blowtorch for the soldering process. Small butane gas blowtorches, using canister gas, can be bought cheaply

from DIY stores. You will also need a roll of solder wire, a small tin of flux and some fine-grade wire wool. Copper pipes and fittings must be perfectly clean before they are soldered. After cleaning with wire wool, the metal is painted with flux to provide a barrier against oxidation until the solder is applied.

Capillary joints

When you look on the shelves of your local DIY store, you will see two types of capillary joint – 'Yorkshire' or integral-ring joints, and end-feed joints. Yorkshire joints are self-soldering, with a small ring of solder already inside the end of the fitting, and just need to be heated with a blowtorch. End-feed joints, on the other hand, do not contain solder and are plain and smooth inside, so you need to apply solder wire to the hot joint. Yorkshire joints are slightly more expensive than end-feed joints. We always buy end-feed joints and have never had any trouble with them.

10 Expert Points

THE FOLLOWING POINTS WILL HELP
YOU TO MAKE A SUCCESSFUL
CAPILLARY JOINT.

1 CLEANING THE JOINT
Whichever type of joint you are using,
make sure it is perfectly clean by polishing
the outside of the pipe and the inside of the
fitting with wire wool.

2 APPLYING FLUX
As soon as you have polished the
surfaces with wire wool, apply a small
amount of flux with your finger. The flux
stops the copper from reacting with the
air while the joint is being made. It is
impossible to make a soldered joint without
flux. It only requires a film of flux over the
surface – this is one situation where you
shouldn't assume that more is better!

3 MARKING
Before soldering, push each pipe into
the joint and use a pencil to mark how far it
goes in, so that if the pipe moves out of the
joint, there is a mark to guide you.

4 PROTECTING THE SURROUNDINGS
Slide a metal sheet, old fire blanket or
soldering mat behind the joint, so that you
don't scorch or burn anything. (At a pinch,
two or three layers of aluminium kitchen foil
could be used instead.)

5 SAFETY
Always wear gloves and goggles
when you're soldering, and make sure that
arms, legs and feet are covered up. Always
have a bucket of water or fire extinguisher
to hand, just in case.

WARNING: Children are fascinated by
soldering, but remember that it is a
dangerous activity. Older children can
watch at a distance – but only if they
are constantly supervised.

6 PREPARING SOLDER FOR AN END-FEED JOINT
Unroll about 20 cm of solder wire before
you start to make an end-feed joint, as it's
difficult to do with a blowtorch in one hand.

7 STARTING THE BLOWTORCH
Turn the blowtorch tap fully on and
light it with a long match, or a lighter with
a long tube between the handle and the
flame. The blowtorch flame should burn
bright blue with no yellowy flame at all.

8 YORKSHIRE JOINTS
Heat the joint with the blowtorch.
Once the flux starts to spit and sizzle, move
the flame up and down the joint to heat it
evenly. It is ready when a small, shiny line
of solder shows where the pipe enters the
joint. Continue heating for a few more
seconds, just so that you are sure it's done
all the way round. Repeat this procedure
for the other end of the fitting.

9 END-FEED JOINTS
For this type of joint, you need to
apply solder when the joint is heated. Start
with a touch of solder at the point where
the pipe enters the fitting. Don't keep the
solder there too long – just long enough to
see if it melts on the pipe. (When the joint
reaches the correct temperature, the solder
will immediately melt on the pipe and flow
into the joint like water.) The solder creeps
into the joint by capillary action. Each joint
only requires about 1 cm or 2 cm of solder
wire. When you can see a fine line of solder
all the way around the joint where the pipe
enters the fitting, it is done.

10 APPLYING EXTRA SOLDER
You can apply an extra dab or two
of solder to both the Yorkshire and end-feed
joints, just to make sure that they are done.
But excess solder will only drip out of the
bottom of the fitting and on to the floor.

Making capillary joints in copper pipe

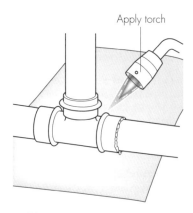
Apply torch

Soldering a Yorkshire capillary joint.

Yorkshire joints

Thoroughly clean the end of the pipe and the inside of the joint with wire wool. Smear a little flux paste inside the joint and on the end of the pipe. Assemble the joint, making sure the pipe is pushed home firmly. Make a small mark with a pencil so that you'll know if the joint moves. Heat the joint and pipe with a blowtorch until you see a small, fine, silver ring of solder appear where the pipe enters the joint. Dab on an extra bit of solder if you think it needs it. Allow the joint to cool fully before moving anything.

End-feed joints

Thoroughly clean the end of the pipe and the inside of the joint with wire wool. Smear a little flux paste inside the joint and on the end of the copper tube. Assemble the joint, making sure that the tube is pushed home firmly. Make a small mark with a pencil so

that you'll know if the joint moves. Heat the joint and pipe with a blowtorch and occasionally dab the solder wire at the point where the pipe enters the fitting. You'll know when the temperature is right because the solder will immediately melt and flow like water. When this temperature is reached, continue to dab the solder into the joint until you see it start to gather at the bottom of the joint, but do not add so much that it drips. Allow the joint to cool fully before moving it.

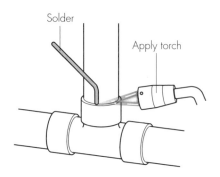
Solder
Apply torch

Soldering an end-feed capillary joint.

Compression joints

Compression joints are available in all forms – bend (or elbow joint), T-joint or straight connector – and are all wonderfully easy to fit. If you only want to do one or two joints, and don't want to go to the expense of buying a blowtorch, compression joints are ideal. All you need are two spanners and a hacksaw or pipe cutter. It is best to buy two 30 cm adjustable spanners, such as those made by Stanley or Bacho, which are available in most DIY stores.

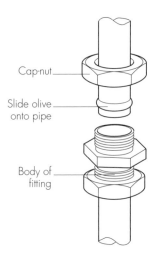

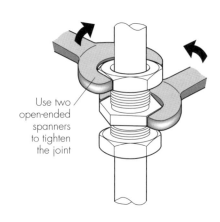

Cap-nut

Slide olive
onto pipe

Body of
fitting

Use two
open-ended
spanners
to tighten
the joint

Fitting a compression joint: Stage 1.

Fitting a compression joint: Stage 2.

The compression joint is made up of the body of the joint, cap-nuts that are screwed on to each part of the joint, and an 'olive' that fits neatly over the pipe and slides under each cap-nut. The olive looks rather like a traditional wide-band wedding ring. As the cap-nuts are tightened, the olive is forced under the body of the joint so that it grips the pipe and forms a watertight seal.

The pipe and olive must be cleaned well with wire wool before you start. You should also inspect the pipe for dents, scratches and anything else that is going to prevent the olive from making a perfect seal, and if you find any of these discard that piece of pipe.

Assembling compression joints

After carefully measuring to allow enough pipe to fully enter the joint, cut the pipe squarely. Use a file to remove any sharp edges, followed by wire wool to polish up both the pipe and olive. Put the cap-nut onto the pipe, followed by the olive. If the olive is of the long-taper type, make sure that the long taper is facing towards the body of the fitting.

Push the pipe into the fitting until it butts hard up against the stop, slip the olive up against the fitting, and then hand-tighten the nut. Check again that the pipe has fully entered into the fitting; mark the position with a pencil so that you will know if it has moved.

To do up the joint, you will need two open-ended spanners of the right size, or two adjustable spanners. Fit one spanner on to the body of the compression fitting. Fit the other spanner on the cap-nut and clench it to compress the olive. Make sure the spanner is a good fit on the nut, or you will spoil both the nut and the spanner. The correct procedure is to

10 Expert Points

THE FOLLOWING POINTS WILL HELP
YOU TO ASSEMBLE COMPRESSION
JOINTS SUCCESSFULLY.

1 PREPARING THE PIPE
Using a hacksaw or pipe cutter, cut the
ends of the pipe squarely. Then, using a flat
file and a round file, remove any burrs from
the cut end before proceeding.

2 THREADING THE CAP-NUT
Put the cap-nut on the pipe, followed
by the olive. Look to see if the sloping
edges of the olive are of equal length. If
one edge is longer than the other, it should
face the fitting (this type of olive is known
as a long-taper olive).

3 JOINT BODY
Push the body of the joint on to the
pipe, so that the pipe end butts up hard
against the ridge inside the fitting. Slip
the olive along the pipe until it is right up
against the fitting. Screw the cap-nut
on tightly by hand, then use the spanners –
one on the cap-nut and one on the fitting –
to clench the joint.

4 TIGHTENING THE JOINT
Usually, you only need to turn the nut
one full turn to make the joint complete. If
you over-tighten, there is a risk that the olive
will be squashed and the joint will leak.If
you are in any doubt about the maximum
turn, it's best to err on the side of caution
and opt for a slightly under-tightened joint.

5 PROBLEMS
If by chance the olive gets crushed, or
you find that the pipe is leaking, use a fine-
bladed hacksaw to cut off the olive. Make
sure you don't cut into the pipe when you're
doing this. Fit a new olive and reassemble
the joint. Make sure that all the component
parts along the way are clean and
absolutely free from grit and other dirt.

6 EXTRA SEAL
There are two ways of adding an
extra seal to a compression joint. You can
use PTFE tape, or a sealing compound that
looks rather like white toothpaste, sometimes
called 'boss white' or 'plumber's mate'.
Although compression joints don't generally
need sealing compounds, if you have got an
awkward one that you can't stop dripping,
a turn of tape or a quick smear of
compound should sort it out.

7 USING PTFE TAPE
If you want to use PTFE tape, just wind
a few turns around the olive before you
assemble the joint. Some plumbers think that
this is bad practice.

8 USING SEALING COMPOUND
To use sealing compound, just smear
a small amount around the olive and then
assemble the joint. Don't be over-generous,
or you'll get the stuff everywhere!

9 CHECKING FOR LEAKS
Once you've made the joint, turn the
water on and watch carefully for leaks. Even
if you only discover a drip every half hour, it
still needs sorting out. Unscrew the joint and
add PTFE tape or sealing compound.

10 DETECTING TINY LEAKS
It is sometimes difficult to know
whether a joint is leaking, as there is
invariably a bit of water splashing around
during the process of assembly. When the
water is back on, wipe the joint carefully
with tissue paper, making sure you get right
into all the corners, and then wipe the whole
thing with another tissue. Make sure that the
pipe is warm, dry and completely free of
grease and condensation, otherwise you won't
be sure that any moisture you find is coming
from the joint. If, after half an hour, you see
that the area is damp, there is definitely a
leak and you will have to repair it.

keep the spanner holding the fitting body stationary while you turn the spanner holding the nut. In most instances, it takes one full revolution of the nut to compress the olive. Marking one face of the nut with a dash of black marker pen will help you to avoid any guesswork since you will be certain that it has moved around one full turn.

Plastic pipes

There are dozens of different types of plastic plumbing pipe on the market. The size of the pipe relates to its outside diameter. Water supply pipes that run through the house or underground use sizes 15 mm, 22 mm and 28 mm. Waste pipes (overflow pipes; washbasin, bath, shower and sink wastewater pipes; soil pipes) come in various sizes. The overflows from toilet systems and tanks in the roof use 21 mm pipes. Washbasin wastewater pipes are generally 32 mm in diameter, while baths and sinks use 40 mm pipes. The soil (sewer) pipes from toilets are 110 mm in diameter.

Advantages

Plastic pipes are wonderful in that they are lightweight and very easy to cut and work with. Plastic technology has advanced to the extent that if you have the right grade of plastic pipe, you can even use them to plumb your central-heating system or supply hot water to taps. Plastic pipes are more frost-resistant than metal pipes, and tend to creak less when they expand underneath the floor. Plastic pipes do not corrode and can be safely connected to copper, steel or lead pipes, without the worry of the metals reacting and causing accelerated corrosion (as would be the case if, for example, you connected a galvanized steel pipe to one made of copper).

Push-fit joints for plastic pipes

Push-fit O-ring joint

To make a joint in a plastic waste pipe, one of the quickest methods (especially suitable for amateur plumbers) is to use an O-ring push-fit fitting. These joints are a little bulkier than the solvent-welded type, but they can be simply pulled apart if things haven't gone quite right! The fittings have a small O-ring in a groove that both seals the joint and grips the pipe. The only thing you have to be careful of is to make sure the end of the pipe is nice and smooth, and lubricated with a smear of soap, otherwise it is possible to damage the O-ring.

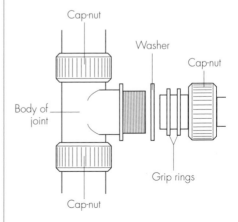

Plastic push-fit supply joint with grip ring.

10 Expert Points

THE FOLLOWING POINTS EXPLAIN THE
BASICS ABOUT PLASTIC PIPES.

1 TYPES OF PLASTIC
Unplasticized polyvinyl chloride
(UPVC) pipes are fairly rigid and are used
for cold-water supply and waste pipes.
Medium density polyethylene (MDPE) pipes
are used for the underground supply to your
house; these are blue or black in colour.
Polybutylene (PB) pipes are used for central-
heating systems and the hot- and cold-water
supply within the house. Polypropylene (PP)
pipes are used for waste pipes (but solvent-
weld joints – see page 52 – cannot be used
on PP pipes). Chlorinated polyvinyl chloride
(cPVC) pipes are suitable for both hot- and
cold-water supply. Cross-linked polyethylene
(PEX) pipes can be used when installing
underfloor heating systems and in other
plumbing situations where it doesn't make
any difference to the efficient functioning of
the system or appliance if the pipes have a
slight tendency to sag.

2 CUTTING PLASTIC PIPES
Plastic pipes are easy to cut with a
fine-bladed hacksaw, but remember that
they must be cut square, so wrap a sheet
of paper tightly around the pipe to give a
square line-of-cut guideline.

3 PIPE CUTTER FOR PLASTIC PIPES
For small-diameter plastic pipes
(22 mm or less), you can get a pipe cutter
that looks rather like a pair of garden
shears. This is a good tool to have at hand
if you're contemplating putting in a central-
heating system using plastic pipes.

4 PREPARING THE PIPE
If you're cutting plastic pipe with a
hacksaw, take extra care to remove burrs
from the inside of the pipe, as they will pick
up dirt – like a dam – and will eventually
form a blockage in the system.

5 PIPE CLIPS
Plastic pipes are a lot more flexible
than copper pipes, so when you're
clipping them to the wall, use twice as
many pipe clips as you would do for a
copper pipe, so that the pipe does not
sag between clips.

6 ADAPTERS
If you haven't got plastic pipes in your
house, but would like to use them at some
time in the future, there are many different
adapters that are designed specifically to
connect copper or steel pipe to any size of
plastic pipe. These can be obtained from
plumbers' merchants (where you can also
get advice if you need it).

7 BUYING ADAPTERS
When buying adapters to fit metal to
plastic pipe, take a small sample of each
pipe with you to ensure the compatibility of
the adapters and pipes.

8 SOLVENT-WELD JOINTS
If you are relatively new to plumbing,
don't use solvent-weld joints in plastic waste
pipes, because once they are in position
you cannot get them apart. Use compression
or push-in fittings instead.

9 PUSH-FIT PIPES
There are two types of push-fit water
supply pipes for cold or hot water. One
type has a metal grab ring, which must be
destroyed to dismantle the pipe, while the
other has a releasable collet (gripper). If
you're a first-time plumber, make sure you
get the collet type.

10 JOINTS
Although plastic plumbing is very
easy to build, the joints are quite large and
bulky. If your plumbing work is going to be
on show, it is best to use copper pipes with
soldered connections.

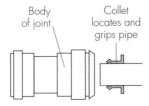

Body of joint / Collet locates and grips pipe

Plastic push-fit joint with collet.

Push-fit collet joint

Push-fit collet joints are very popular, especially for connecting taps to copper pipe with a short, flexible connector. The collet is like a one-way gripper. You push the pipe in, but you can't pull it out. An O-ring, built into the fitting, seals the joint.

The end of the copper pipe must be free of burrs, or the O-ring may get damaged. To produce a nice, smooth end on a pipe when you intend to use push-in joints, it is best to use a pipe cutter rather than a hacksaw to cut it.

If you need to dismantle the joint, you will find a collar poking out of the joint which, when pushed and held in, allows you to release the pipe from the collet. To achieve this, you need to hold the collar in and pull the pipe at the same time. There may be a small semicircular clip in a groove on the collar: this is a locking device that should be removed if you want to dismantle the joint.

One-way grip-ring joint

The other common type of push-fit joint uses a one-way grip ring (or grab ring). This works in a similar way to a collet joint, but cannot be dismantled without unscrewing the end and crushing the grip ring with pliers.

To make the joint, the cut pipe is lubricated with silicone grease or a little soap and pushed home firmly. It is important that the pipe is squarely cut; if the tube is plastic, an internal support should be inserted into the end of the pipe. To re-use the joint, the end is unscrewed and a new grip ring fitted, with the little teeth facing towards the joint. (When you dismantle push-fit joints, it is a good idea to lay out the component parts in their proper sequence so that you know how they go back together.)

Cutting and bending plastic pipes

Plastic pipes are easy to cut and bend without specialist tools.

Cutting

For cutting plastic supply pipes, you need only a fine-bladed hacksaw (sometimes called a junior hacksaw) with more than 120 teeth per inch. When cutting, the most important factor is to ensure that the cut end of the pipe is square. To do this, wrap a sheet of paper around the pipe and roll it tight, then use a pencil to mark the line of the edge of the paper. Keep the pipe steady during sawing by putting it across a couple of kitchen chairs or a workbench, and saw gently. Use a flat file to remove the burrs from the inside and outside of the cut end.

For small-diameter plastic pipes (22 mm or less), you can get a pipe cutter that looks rather like a pair of garden shears. Waste pipes (32 mm and larger) are cut with a junior hacksaw in the same way.

Bending

Hard plastic supply pipes can be bent by heating them slightly on the outside with a blowtorch. This has to be done with extreme care so that you don't burn through the pipe. Wear gloves and gently bend the heated pipe into a curve, holding it in position until it is cool (it sometimes helps to do this on a flat surface as you may be there for several minutes waiting for the pipe to cool).

Flexible plastic supply pipes can be bent using special angled metal fittings (sometimes called corner clamps) that are screwed to the wall: the pipe is forced into the fitting. Flexible supply pipes can be bent 'cold' in this way to a maximum radius of eight times the pipe diameter, so a 15 mm supply pipe can be bent to a radius of 120 mm.

Waste pipes (size 32 mm and larger) cannot be bent by hand – you need to use angled joints.

Plastic waste pipes and joints

The introduction of plastic has revolutionized waste pipe systems (overflow pipes; washbasin, bath, shower and sink wastewater pipes;

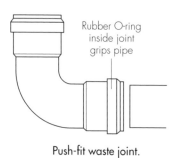

Rubber O-ring inside joint grips pipe

Push-fit waste joint.

soil pipes). The secret of building an efficient, long-lasting, easy-to-maintain waste system is to choose the correct type of plastic for the task, and to make sure that the pipes, joints, solvents and lubricants are compatible with each other. The Expert Points on page 51 will enable you to make an informed choice of materials.

Making a push-fit joint in a plastic waste pipe

Push-fit joints are good on three counts: they are wonderfully easy to fit, they are slim and unobtrusive, and they are easy to dismantle if you make a mistake and have to start again.

Method

1   As with all plumbing, it is important that the pipe is cut squarely at the end. Use a fine-bladed hacksaw to make the cut.

2   Remove the sharp edge from the cut pipe with a file, so that it is slightly tapered. This taper allows you to easily push the pipe past the seal in the joint.

3   Wipe a bit of silicone grease on the pipe and carefully push it into the joint, making sure that you're not entering it at an angle.

Curing a leak

If you have a small leak on a push-fit waste pipe, this will usually have happened because the pipe and joint are not in alignment. This may happen if you force the pipe to one side or up and down, either because one of the pipes is not the right length, or you've used the wrong joint.

10 Expert Points

A RUNDOWN OF USEFUL INFORMATION ABOUT PLASTIC WASTE PIPES.

1 TYPES OF PLASTIC FOR WASTE PIPES
Unplasticized polyvinyl chloride (UPVC), polypropylene (PP), modified unplasticized polyvinyl chloride (MuPVC), and one or two other plastics are used for waste pipes.

2 PROPERTIES OF PLASTICS
Each type of plastic has unique characteristics and qualities that make it ideal for a specific use. One plastic might have the advantages of availability in continuous lengths and resistance to freezing, and the disadvantage that it sags and expands when it gets hot, and cannot readily be joined with a solvent-weld joint. To correctly match each type of plastic pipe to the appropriate solvents, lubricants and joints, read the manufacturer's literature and follow the assembly guidelines very carefully.

3 FIXING WASTE PIPES
When you are fixing the pipework to walls, or hanging it under a wooden floor, be generous with the fixing clips so that the pipe doesn't sag along its length.

4 CHANGING EXISTING WASTE SYSTEMS
If you want to make changes to an existing pipe run – perhaps make it longer, or cut in to fit a new sink – it is sometimes easier to scrap the old system and replace it.

5 COMPRESSION JOINTS FOR WASTE PIPES
Compression joints are made in a variety of plastics, and in all manner of sizes, angles, junctions and forms, to fit matching pipes. To fit a compression joint, the pipe is prepared, pushed into the joint, and then the cap-nut is tightened so that a captured rubber O-ring grabs the pipe.

6 PUSH-FIT JOINTS FOR WASTE PIPES
Push-fit waste joints and pipes are manufactured in groups or families from the same type of plastic, in all manner of sizes, angles, junctions and forms.

7 SOLVENT-WELD JOINTS FOR WASTE PIPES
Solvent-weld joints and pipes are manufactured in groups or families from the same type of plastic, in all manner of sizes, angles, junctions and forms. Solvent-weld joints are designed so that the 'female' connection receives the 'male' pipe. After the pipe is cut and prepared, one (or both) of the mating faces is brushed with solvent, and then the pipe is slid into place inside the joint. The solvent fuses the joint and pipe together. This type of joint is commonly used for waste pipes that are hidden from view and not easily accessible when you want to be absolutely certain that the joints are going to hold.

8 EXTENDING EXISTING SYSTEMS
If you intend to extend an existing system with solvent-weld pipes and joints, it's a good idea to use a compression joint to link the new system to the old, just in case the old system cannot be solvent-welded. WARNING: Solvents are variously inflammable, toxic when the fumes are drawn through a cigarette, and generally unpleasant. Always follow the manufacturer's safety guidelines.

9 T-JOINTS (OR TEES)
When you are building a waste system that uses swept tees and branch joints, make sure that the arrows printed or stamped on the joints are pointing in the direction of the flow.

10 COMPRESSION JOINTS
When you are tightening up compression joints, be very careful that you don't cross the threads.

If one of the pipes is the wrong length, dismantle the pipework and fit a shorter or longer pipe as necessary. If the pipe is no more than a few millimetres short, you can make the adjustment simply by pulling the pipe slightly out of the joint at each end.

If the above is not the cause of the leaking joint, it may be that the seal was 'nipped' when you assembled the joint. If this is the case, dismantle the pipework and fit a new joint.

Allowing for expansion

All plastic pipework expands when hot water is run through it. To allow for this expansion, you need to make some adjustments. With the pipe pushed hard into the joint, make a small pencil mark on the pipe. Carefully wiggle the pipe out of the joint by about 10 mm. Repeat this process on each joint in the waste pipe system. This ensures that the pipes will not bend and creak when hot water runs through them. (To allow for these adjustments when initially cutting a length of the pipe, measure the distance required to push the pipe fully into the push-fit joint(s), and then subtract 10 mm for each joint.) That said, the technology is moving so fast, with all manner of new products and fixtures coming on the market, that you have to make doubly sure that you are using the correct products and procedures for the plumbing task that you are carrying out. If you are in any doubt, ask the supplier for product advisory sheets before you buy any materials for the work you need to do.

Making a solvent-welded joint in a plastic waste pipe

One of the simplest methods for connecting waste pipes that are 32 mm or 40 mm in diameter is to use solvent-welded pipework. The pipes and fittings are generally made of UPVC or MuPVC and are white in colour. Solvent-welded fittings are quite slim and tidy, taking up the least space of all waste-pipe joints.

WARNING: It is most important that you do not smoke while making solvent-weld joints, because the vapour given off by the solvent can be extremely harmful when drawn through a cigarette. The solvent is also highly flammable and must be kept in a secure cupboard well away from children. Once you have completed your plumbing work, we recommend that you dispose of the solvent container. Remember that you cannot solvent-weld polypropylene pipes. The order of work is as follows.

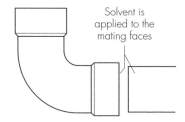

Solvent is applied to the mating faces

A solvent-weld joint used for supply and waste pipes.

Method

1   Wrap paper around the pipe to give a square guideline, and make the cut with a fine-bladed hacksaw.

2   Use a flat file to remove the burr

on the inside and outside of the cut end. Use a small piece of sandpaper to slightly roughen the mating faces – the inside of the joint, and the outside of the pipe. This roughened surface provides a 'key' for the solvent.

3　Prior to using the solvent, have a dry dummy run. Push the pipe into the joint and mark a line on the pipe with a pencil to show where the joint ends (this will also indicate the area to be glued). Also make a mark on both the pipe and the joint so that you know how they should be aligned when you come to reassemble them.

4　Some solvent manufacturers recommend cleaning the pipe with a chemical cleaner before applying the solvent cement. If this is the case, use a clean cotton rag and a little of the cleaner to wipe the pipe and the joint.

5　The solvent is usually supplied with a small brush. Use this brush to paint the inside of the joint and the outside of the pipe, then immediately push the pipe into the joint and twist it around until the pencil guide marks line up. In no more than a couple of minutes the solvent will have welded the two plastic surfaces together – forever! Allow a few hours for the solvent to cure fully before running hot water through the pipework.

6　Solvent-weld joints do not move when the plastic pipe expands. If you have a long length of waste pipe (more than 3 m), incorporate a push-fit joint somewhere in the

system to allow for expansion. You can buy special expansion joints that have a solvent-welded socket at one end and a push-fit end at the other. The pipe inserted into the push-fit end should only be 20 mm past the seal, in order to allow sufficient room for the pipe to expand when it gets hot.

Making a compression joint in plastic waste pipe

Compression joints are a little bulkier than the push-fit type, but this doesn't matter if the pipework is to be boxed in. Each end of a compression joint consists of a large cap-nut, a thin plastic washer and a tapered rubber seal. (Note that some compression joints do not have the thin washer.)

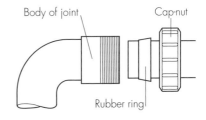

Plastic compression joint used for waste pipe.

Method

1　Cut the pipes squarely by using a wrap of straight-edged paper as a guide. Use a fine-bladed hacksaw to make the cut, working with a rapid, light stroke. If you push too hard, the blade will break or jam.

2　Use a file to remove burrs from the inside and outside of the cut ends.

3　Unscrew the large cap-nut from the

end of the joint and slide it over the pipe, followed by the plastic washer and the rubber seal. Make sure that the thin, tapered end of the rubber seal faces the joint. Some compression joints have symmetrical seals – these can go either way round.

4　Push the pipe into the joint and slide the rubber seal up the pipe until it is in the end of the joint. Slip the plastic backing washer up the pipe and screw the nut home until the pipe is secure.

5　Make a pencil mark where the pipe enters the joint, then pull the pipe out by 10 mm to allow for expansion when hot water goes through it. Screw up the nut as firmly as you can by hand. Do not use a spanner or wrench.

6　The occasional small leaks that occur on compression joints in waste pipes are usually caused by some misalignment of the pipe within the joint. If you find that for some reason or other you are pushing the pipe into the joint at a skewed angle, there is a likelihood that the joint will leak. If a pipe does leak you will probably need to adjust the length of the pipe. You can ease the pipe in or out of a compression joint simply by loosening the cap-nut and moving the pipe to suit. Don't move the pipe more than 5 mm over and above the initial 10 mm that you added for expansion.

Cold-water storage tank

The cold-water storage system usually consists of a large tank in the loft or roof space, or sometimes a large tank on a flat roof. If you are a home owner you will, sooner or later, need to familiarize yourself with the workings of the storage tank – its location and the way the valve, overflow and distribution pipes work.

In essence, the primary function of the cold-water storage tank or cistern is pretty straightforward. However, over the years its recommended location, size, structure and function have evolved to suit changing needs. You might have two tanks, a single small plastic tank, or sometimes no tank at all. All this makes for a slightly confusing scenario! The following Expert Points address some common queries, and have been put in question and answer form.

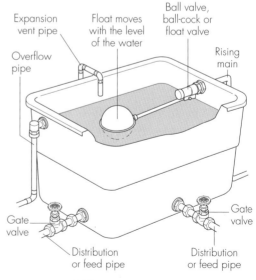

Cold water storage cistern.

10 Expert Points

1 Why haven't I got a storage tank? In some old houses, and in many new houses, there is a direct water system that uses water direct from the mains. Everything in the house – hot water, cold water, appliances – is fed directly from the mains supply. Some direct water systems do not include a cold-water storage tank. (See page 26.)

2 If technology has made it possible to do without a cold-water storage tank, why not do without? Most people prefer to have a storage system for the simple reason that it provides an emergency supply to fall back on if the mains water is cut off.

3 There are two tanks in my loft – a galvanized one that is empty, and a plastic one that is full. Why? The likelihood is that the galvanized one is disused. It has been left in the loft simply because it is too big to go through the hatchway. Either ignore it, or use an angle grinder to cut it into movable pieces and remove it. The plastic tank is the one to focus your attention on.

4 There are two plastic tanks in my loft – a large one, and a smaller one at a higher level. Why? The large tank is the cold-water storage tank. The small tank (known as a feed-and-expansion tank) keeps the hot-water system topped up, and collects water when the hot-water system overheats.

5 What is the function of the storage tank and why is it sometimes described as an 'indirect storage tank'? Mains water rises under pressure to the storage tank, where it sits until it is drawn off to feed the toilets, all the taps in the house except the cold-water tap in the kitchen, or one of your appliances.

6 My storage tank is ancient, and might even be made of lead. Is this a problem? If your tank is made of lead, copper or galvanized steel, it poses a problem. Lead is poisonous, and copper and galvanized steel corrode. It is best to replace the tank with a new plastic one. (A lead tank can be used in the garden as a decorative plant container.)

7 I want to fit a new plastic storage tank, but the hatchway to the loft is very small. Will I need to cut a new hatchway to the loft? Not necessarily. It is possible to buy a plastic tank that folds up, so it can be passed through the existing hatchway. Make sure that you get one with a tight-fitting lid and side stiffeners.

8 I have found feathers in my storage tank, but as our drinking water comes straight from the mains supply, does it matter? Your drinking water comes straight from the mains via the cold tap in the kitchen, but what about when you clean your teeth in the bathroom? That water comes from the storage tank. Feathers and droppings from birds, bats or mice that have got into the loft are a health hazard. If your storage tank is not properly covered, buy a new tank. New storage tanks comply with modern by-laws that require tanks to come complete with a close-fitting lid that excludes just about everything that might get into the water.

9 I am going away on a long winter holiday – do I need to empty the storage tank? If you want to avoid the possibility of damage if there is a leak or pipe burst, the answer is yes. Turn the water off at the mains, drain the tank, open the hatchway to the loft and leave a tried and trusted sealed-tube electric convector heater on tick-over.

My plumber tells me that my new tank must comply with By-law 30. What is he talking about and is it expensive? The new by-law is concerned with keeping water clean. The new tank will have a lid, and filters on the various pipes. To assess the cost, phone a plumbers' merchant and ask for the list price of a 225-litre storage tank. Find out the cost of a kit that comes complete with all the fixtures and fittings, and the cost of buying the various items individually. This will enable you to establish the cheapest way to buy the materials.

WC cisterns

In many ways, the cold-water storage tank in the loft and the WC cistern in the toilet are much the same in function and design. The level of the water is controlled by a float valve, there is a feed pipe for the water to go out, and there is an overflow pipe that comes into play if the valve fails.

When the water level in the cistern falls, the ball float drops and opens the valve; when the water level rises, the ball float rises and closes the valve. When the flush is pressed or pulled, the water in the inverted-U-shaped siphon in the body of the cistern is lifted to the point where the water siphons off from the cistern to flush the WC pan. And, of course, when the cistern is flushed, the water level drops within the cistern, the ball float falls, and the valve opens to allow more water into the cistern.

WC cisterns generally last for a long time; however, either the float valve or the flap valve within the siphon mechanism may fail from time to time, which means that your cistern won't work. The good news is that the design of a WC cistern is so basic that it is easy to service or replace the valves when necessary.

Portsmouth-type float valve

The commonest type of ball or float valve in the UK is known as the Portsmouth piston. Made of brass and virtually indestructible, the valve is opened and closed by means of a cylindrical plug that moves backwards and forwards within a horizontal chamber. When the ball float is in the 'up' position, the other end of the float arm pushes the plug along its chamber, with the result that a rubber disc at the end of the plug is pressed hard against a water inlet nozzle, cutting off the water. When the ball float is in the 'down' position, the float arm pulls the plug back along its chamber, with the result that the rubber disc located at the end of the plug eases away from the water inlet nozzle, allowing water into the cistern.

A Croydon valve operates along much the same lines, with the difference that the plug moves vertically instead of horizontally. The servicing of both types of valve involves cleaning the piston and/or replacing the rubber disc. While the Portsmouth valve and the much older Croydon valve are both available, modern diaphragm valves are in many ways a more efficient option.

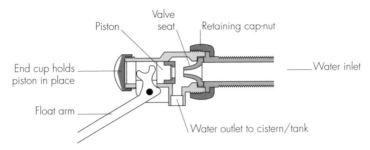

Portsmouth-pattern ball float valve (cross-section).

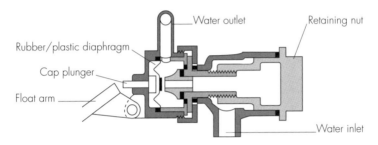

Diaphragm-type ball float valve (cross-section).

Diaphragm-type float valve

The modern diaphragm ball or float valve is made primarily of plastic and rubber. In action it is much the same as the traditional brass valve – with the float being lifted up and down by the level of the water. But when it comes to the business end of the arm, the working action is such that when the water rises, a small plastic piston pushes against a rubber window or diaphragm, with the effect that a rubber disc on the other side of the diaphragm pushes against the water inlet nozzle to cut off the water. And of course, when the water level drops, the float falls, with the effect that the other end of the arm ceases to press against the diaphragm and the water flows out of the nozzle.

U-bends and traps

U-bends or traps are a means by which a pocket of water is held within a waste pipe in order to stop smells and gas from the drains getting back into the house. They feature in all waste pipe systems (overflow pipes; washbasin, bath, shower and sink wastewater pipes; soil pipes).

The early lead and copper traps in sink and bath waste pipes were just a U-shaped bend in the pipe, and some modern plastic waste pipes are still in that form. However, the design has now evolved and the essential U-shape has been miniaturized and is contained within an easy-to-fit, plastic, bottle-shaped unit – which is suitable for all sink, washbasin, bath and shower waste pipes.

10 Expert Points

IDENTIFY THE MOST APPROPRIATE TRAP FROM THIS LIST.

1 TRADITIONAL LEAD U-BEND TRAP
This unit is merely a length of lead pipe that has been fashioned into a double U-bend, with a small screw-in access cap fitted in the bottom of the 'U'. To clean the trap, you unscrew the access cap and remove the sludge. However, the very presence of a lead trap is a concern – if you still have a lead trap on one of the sinks, are there other lead pipes in the house? If so, they need to be removed.

2 TRADITIONAL COPPER OR BRASS U-BEND TRAP
This unit is fine, as long as you can undo the access cap. The moment you have problems, it is best to replace it with a modern plastic fitting.

3 TUBULAR S-TRAP OR DOUBLE-U-TRAP (PLASTIC)
This is very much like the old lead or copper traps. It is a good, low-cost solution for a kitchen sink or a large bathroom washbasin that has a large outflow and plenty of room for fitting and servicing, and where the need is for the waste pipe to travel in a downwards direction. The swivel joint at the centre of the 'S' allows you to adjust the trap for a whole range of situations and locations. Cleaning is achieved simply by unscrewing the unit.

4 TUBULAR TELESCOPIC TRAP (PLASTIC)
This is a good choice for a sink or washbasin that has a large outflow, and for a location where you know that the waste pipe needs to travel horizontally back along or through a wall, but you aren't sure of the precise height. The horizontal linkage and the telescopic vertical pipe will allow you to adjust the pipe to suit an infinite number of possibilities.

5 TUBULAR J-TRAP (PLASTIC)
This is a good solution for a washing machine outlet, where the inlet pipe needs to be at a high level, and the outlet pipe needs to be positioned low down and running out in a horizontal direction.

6 BATH TRAP WITH INTEGRAL OVERFLOW (PLASTIC)
This unit is designed specifically for bath and sink applications, when there is a need for length and flexibility in the overflow pipe. The design is such that you can easily adjust the length and angle of the overflow pipe to suit your chosen bath or sink. There are many different designs, including cleaning caps and U-bend forms, to suit just about every possible application. It is best to postpone getting this unit until after your bath or sink is in place.

7 LOW-LEVEL SHOWER TRAP (PLASTIC)
This unit has been designed specifically for showers – where there is a small outflow, and where there is limited room between the underside of the shower tray and the floor. This unit allows you, for example, to set a shower tray directly on a concrete floor, and to fit the pipework without the need to chop into the concrete. Because the U-bend is permanently built in and hidden from view, the design allows cleaning to be achieved by lifting the top grid and removing a little tubular sludge-catcher. Some models of shower trap are very fragile, so buy the best that you can find.

8 STANDARD BOTTLE TRAP (PLASTIC)
Designed for the average small washbasin where there is likely to be a small outflow, this unit has a horizontal outflow pipe pointing to one side. The bottle shape makes it a good choice for a small washbasin where there is a shortage of space. To remove sludge, you simply unscrew the bottom half of the bottle.

9 TELESCOPIC BOTTLE TRAP (PLASTIC)

This is the same design as the standard bottle trap, with the addition of a telescopic extension inlet tube coming up from the top of the trap. The telescopic bottle trap outlet points to the side – for a horizontal fitting – although there is an optional unit that can turn the outlet into down-pointing mode.

10 ANTI-SIPHON TRAP (PLASTIC)

This trap is designed specifically for situations where the outflow ranges between light and heavy. There is a small valve installed on the top of the trap which allows air to enter if the 'siphon' tries to occur. This is called a 'siphon break'.

There are lots of different U-traps on the market, each suitable for a particular purpose, so you need to get the right one. U-traps are generally just screwed in place.

The Expert Points here are in the form of a miniature glossary that will explain precisely which type of trap you already have, or which type you are going to need.

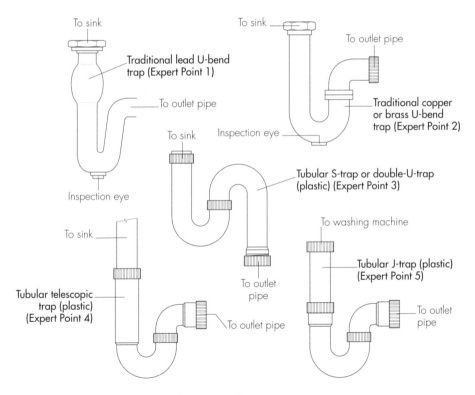

U-bends and traps (*see also* Expert Points opposite).

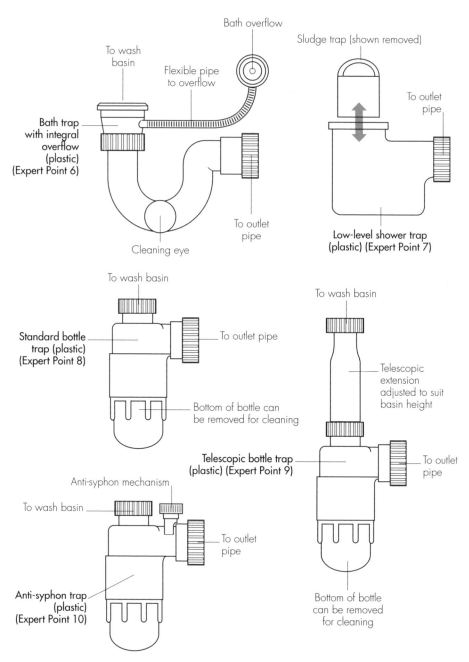

To wash
basin

Bath overflow

Sludge trap (shown removed)

Flexible pipe
to overflow

To outlet
pipe

Bath trap
with integral
overflow
(plastic)
(Expert Point 6)

Cleaning eye

To outlet
pipe

Low-level shower trap
(plastic) (Expert Point 7)

To wash basin

Standard bottle
trap (plastic)
(Expert Point 8)

To outlet pipe

Bottom of bottle can
be removed for cleaning

To wash basin

Telescopic
extension
adjusted to suit
basin height

Telescopic bottle trap
(plastic) (Expert Point 9)

To outlet
pipe

Anti-syphon mechanism

To wash basin

To outlet
pipe

Anti-syphon trap
(plastic)
(Expert Point 10)

Bottom of bottle
can be removed
for cleaning

Modern plastic bottle traps and U-bends for specific uses (*see also* Expert Points pages 58–59).

Gullies, inspection chambers and drains

Many people become very confused when it comes to understanding the definition of, and the difference between, a gully, a drain and an inspection chamber. Of course, this confusion doesn't matter too much in the normal course of events, but when it comes to trying to explain a problem to a plumber over the phone, or trying to order materials from a plumbers' merchant, understanding and clarity are everything.

Gullies

A gully is the grid-covered, water-filled trap – like a U-bend – that you can see at the open end of a drainage system. Or put another way, it's the bit somewhere outside the kitchen that takes the wastewater from the kitchen sink, or the bit that you see at the

bottom end of a rainwater pipe (and which keeps getting blocked with leaves). All gullies need servicing, but it is the kitchen waste gully that you need to keep a special watch on, because it is prone to getting blocked by the remnants of food, etc.

In older houses, there is often a two-pipe drainage system. Wastewater runs down one pipe system, and sewage from the toilet runs down a dedicated soil pipe. The wastewater from the upstairs bath tub and bathroom washbasin runs into a funnel-shaped hopper (usually seen on the outside wall somewhere near the bathroom window), down a waste pipe and into a grid-covered gully. There it is joined by wastewater from the kitchen sink, and moves on through an inspection chamber and into the sewer. The waste from the toilet runs down its own dedicated

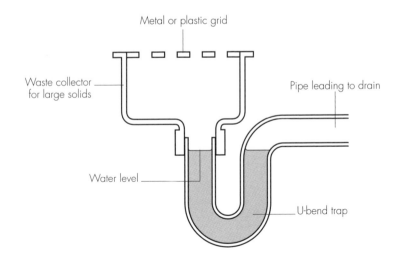

Metal or plastic grid

Waste collector for large solids

Pipe leading to drain

Water level

U-bend trap

A yard gully (cross-section).

stack or soil pipe straight into the inspection chamber, where it is washed on its way by the wastewater from the bathroom and kitchen.

In newer houses, there is usually a single-stack system. The waste from the baths, washbasins, kitchen sink and toilets runs down the same large-diameter pipe. (Sometimes the water from the kitchen sink discharges directly into its own trapped gully.)

Unblocking a gully

Gullies are easy to unblock. Put on the longest rubber gloves that you can find, lift out the grid and start scooping out the smelly detritus. (If you eat a lot of fatty food, are lazy when cleaning out your teapot, or are in the habit of washing out buckets of this and that in the kitchen sink, they will all come back to haunt you when you clean out the gully!) However, you can easily get to all the mess before it gets a chance to block your drains. When you have cleared the solids, it's a good idea to blast water through the grid and the gully with a garden hose. Finally, wash your gloved hands and the area in and around the gully with disinfectant.

Drains and inspection chambers

All the underground pipes of the waste system are known as drains, but the access holes that link the straight runs of underground pipe (or that mark the points where the straight runs change direction) are known as inspection chambers or manholes. The thinking behind this design feature is that if there is going to be a blockage in the drains, the likelihood is that it will occur at one of the bends or junction points of the drain.

When you lift the lid of an inspection chamber, you should see a nice, clean, straight or Y-shaped half-channel running across the base. However, if you discover that the

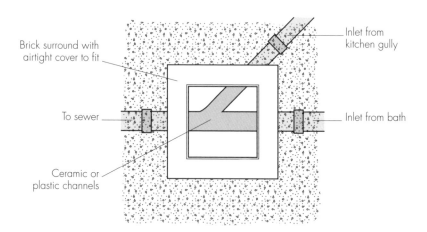

Brick surround with airtight cover to fit

Inlet from kitchen gully

To sewer

Inlet from bath

Ceramic or plastic channels

An inspection chamber (plan view).

channel is full of solids, or worse still the whole manhole is flooded, the drain is deemed to be blocked. Since a blocked drain is a potential health risk, some local councils will send out a team to clean the system free of charge – or at least at a minimal cost. It's always a good idea to try this as your first course of action. And, of course, if you are elderly or infirm, most councils will be only too happy to help and advise.

Unblocking a drain

Unblocking and cleaning a drain is easy, but messy and somewhat hard on the nostrils. It is best to wear old clothes that you are prepared to throw away afterwards, complete with heavy-duty rubber gloves and full-length boots.

Take your set of drain rods and ease the first rod into the drain. Screw the second rod to the first, and push the two rods forward with a clockwise screwing action. Fit the third rod to the second, and so on. When you feel an obstruction, move the rods backwards and forwards with a gentle but firm joggling action. When the glorious moment comes when you hear a gurgling sound and the chamber clears, you simply pull the rods out, still screwing them in a clockwise direction. To finish, hose everything down – the hole, rods, gloves and boots. Put the clean rods to one side to dry out. Turn on all the cold-water taps and flush the toilet. Finally, throw all the clothes you have been wearing while carrying out the work in the dustbin, and then have a shower or bath. While it is a relatively easy procedure, the proviso must always be: only do it if the rest of the household knows what you are doing.

WARNING: Cleaning drains is a relatively safe and easy activity, as long as you DO NOT do any of the following things during the procedure: smoke, eat, poke your ears, eyes or mouth with a sludge-covered finger, or trip over and fall down the hole. Children can watch at a safe distance – it's a wonderfully educational process that will fascinate them – but they cannot be allowed to help as it's a potentially dangerous operation. If your children want to watch, then you need to be ready with answers to their inevitable questions on the whys and wherefores of the contents of the drain and how they got there.

Cesspits and septic tanks

If you live in a city, town or large village, the sewage from the toilets, and the wastewater from the baths, sinks, basins and washing machines, is directed through an underground pipe and inspection chambers and on into the public sewage system. In country areas, however, where houses are remote from public sewers, the waste runs into either a cesspit or a septic tank. If you are thinking of buying a house in the country, or if you already live in a remote property and are thinking of making changes to your plumbing, it is vital to know just what type of system you are dealing with. The Expert Points overleaf will provide the answers.

10 Expert Points

FIND OUT ALL YOU NEED TO KNOW
ABOUT CESSPITS AND SEPTIC TANKS.

1 IDENTIFYING A CESSPIT SYSTEM
A cesspit is a large brick- or concrete-lined pit that acts as a holding chamber for your waste. Usually, there will be an inspection chamber before the pit, a ventilator on the top of the pit, and that's just about it. The trouble with a cesspit is that it is smelly and constantly needs emptying. If you are buying an old house in the country, don't be led into believing that a cesspit and septic tank are one and the same thing – they aren't. Generally speaking, a cesspit is bad news. In action, the waste goes through the inspection chamber and into the cesspit, where it waits until the pit is emptied. One of the drawbacks of a cesspit is that you have to pay for it to be emptied.

2 EMPTYING A CESSPIT
When does a cesspit need emptying? The smell will let you know. If you see any ooze, contact the local council. Sometimes they will help, either by emptying the cesspit or giving you a contact list of firms who will undertake the job for you.

3 UNACCEPTABLE PRACTICES
Traditionally, cesspits were 'improved' simply by emptying them and knocking a drainage hole at a point well below ground level. While this undoubtedly worked in the past – especially on sandy and chalky soils – nowadays the planning authorities would not be too pleased if you were to do this.

4 CONVERTING A CESSPIT
You can opt to have your cesspit converted into a septic tank, especially if you have a large property with no neighbours nearby. While this isn't a particularly difficult procedure, it does involve a great deal of digging. So, to save yourself time and trouble, it is probably best to employ a specialist firm.

5 COVERINGS ON CESSPITS
Be wary if your cesspit has a covering of railway sleepers – they might be rotten. If they are, it is best to replace them with concrete sleepers or new wooden sleepers.

6 SEPTIC TANK KITS
It is now possible to buy a kit for a completely self-contained septic tank, complete with its own sludge turner, filter bed and so on. There are many different products on the market, ranging from massive self-contained units, through to small eco-friendly units that turn the sludge into compost. It is best to research all the possibilities before committing yourself. Take into account your needs: don't, for example, buy a system that will turn the sludge into compost unless you can find a use for the compost that will be produced.

7 IDENTIFYING A SEPTIC TANK SYSTEM
A septic tank system has an inspection chamber, a large tank or chamber, and a filtration system. The waste goes through the inspection chamber and into the large tank, where the solids fall to the bottom and the water flows off through the filtration system and on into the soil.

8 ADVANTAGE OF A SEPTIC TANK
A good septic tank hardly ever needs emptying. If it doesn't smell and the topsoil isn't oozing, leave it alone.

9 SEPTIC TANK MAINTENANCE
If the septic tank begins to smell, have it emptied and either wash or replace the layers of sand and gravel in the filter bed.

10 CRUSTY WATER
Don't be alarmed if you open your septic tank and see a large crust over the surface of the water. It shows that the system is working. Leave it until it starts to smell or you notice that the tank is overflowing.

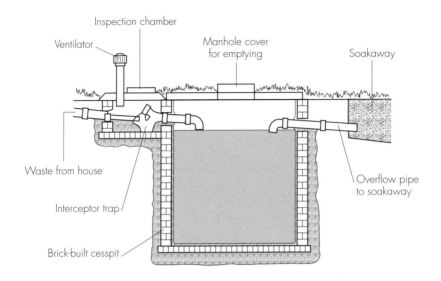

A traditional cesspit system.

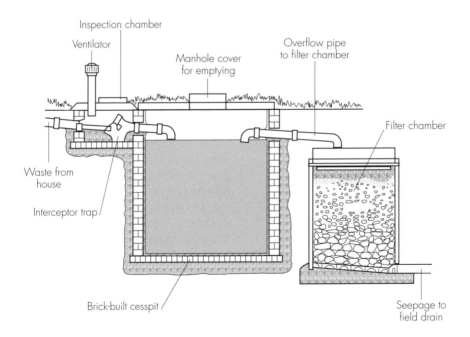

A pre-1960s septic tank.

CHAPTER 5

General plumbing projects

DIY versus a plumber

When it comes to plumbing – the sudden unexpected emergency, or new changes that you want to make – the big question is do you want to do it yourself, or are you going to bring in a plumber? Well, just like every other task in life, much depends on your age, your physical strength, the time available, your confidence and the size of your bank balance. If we take it that you are strong and have enough time to do the job, the most important factor is your confidence. If it is late at night, and you are tired and under stress, it's not a good idea to try to mend something like a burst pipe. But if you are relaxed enough to turn off the mains and wait until the next morning, when you have had a sound night's sleep and a good breakfast, it's fine to do it. And much the same goes for any of the other projects. A keen DIY enthusiast who is confident about approaching the task and has the right tools can easily manage most plumbing projects. The same applies to the bigger projects such as fitting a central-heating system. The only thing that you MUST NEVER DO is work on the gas boiler. All work to the gas boiler must be carried out by a CORGI-registered engineer. If you notice an odd smell or anything else that is unusual, shut the gas down and call in a specialist.

And now for the exciting bit!

New cold-water storage tank in the loft

If your cold-water storage tank is obviously very old – made of galvanized steel – or the old plastic cistern has sprung a leak, it is time to fit a new one. Make preparations by asking a friend to help, assembling all your tools, and making sure that the loft is well lit. Purchase the new tank complete with float valve and tank plate; plastic pipes, compression

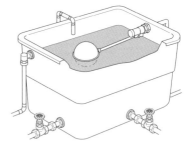

Cold water storage tank (see page 54).

joints and connector to link up to the existing overflow pipe; copper pipes, compression joints and connectors to link the float valve to the existing rising-main pipe; copper pipes, connectors, compression joints and gate valves to link the new tank to the existing distribution pipes. Now you are ready to start: put your work clothes on. Warn the household and then the fun can begin.

Method

1   Start by turning off all your water-heating appliances. Then turn on the cold-water taps in the bathroom, and turn off the stopcock on the rising main.

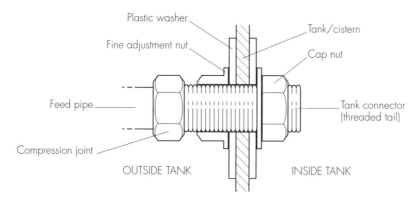

A compression joint for a new cold water cistern.

2   When the water stops running out of the bathroom taps, go up into the loft with a bucket and mop, and remove the last dregs of water from the old tank. Use a spanner to unfasten all the connections to the old storage tank.

3   If the tank is galvanized, drag it to another part of the loft and slide it on to a sheet of plywood or an old door (ensuring that the load is evenly distributed).

4   Check that the old tank was sitting on a good, solid base or a sheet of plywood (25 mm thick), and set the new tank in position. Align all the existing pipes for best fit.

5   Starting with the distribution pipes, modify and extend the existing pipes, drill holes of an appropriate size at a point about 60 mm above the bottom of the tank, and then connect up.

6   The order of connection is as follows. Slide a plastic washer on to the threaded tail of the tank connector. From inside the tank, slide the threaded tail through the hole to the outside. Slide the second washer on to the tail, wrap PTFE tape around the thread, and then screw on the nut and clench tightly with a pair of spanners.

7   Next, fit gate valves to the existing distribution pipes, and then link up with the connectors that you have just fitted. Use the spanners to clench the compression joints.

8   The overflow pipe is connected in much the same way as the distribution pipes, the only difference being, of course, that the connection is made about 100 mm down from the rim of the tank, and the pipes and fittings are made of plastic rather than copper.

9   When you come to fitting the float valve, start by drilling the hole at a point about 75 mm down from the rim of the tank, so that it is about 25 mm higher than the overflow hole. The order of work from now on is as described below.

10 Slide a plastic washer on to the threaded tail of the valve and push the tail through the hole. Slide the

tank support plate on to the threaded tails, followed by the second plastic washer and the fixing nut. Finally, connect the existing rising main pipe to the float valve.

11 When you are happy with all the connections, slide the expansion or vent pipe through the hole in the tank lid, and call down to your helper to turn the stopcock on. When the water begins flowing out of the bathroom taps, turn the taps off and wait for the tank to fill up. It is important to keep a constant watch while the level of the water in the tank rises above the various holes and fittings.

12 Lastly, when the level of water within the tank reaches the top, adjust the float arm so that, at valve close-off point, the water level is lower than the overflow pipe. The job is now done.

Replacing a toilet

It is a relatively easy task to replace an existing toilet, or install a completely new system. However, you do have to make sure that the WC cistern, WC pan and link-up pipes are perfectly compatible. The secret of success is forward planning. You must have a clear understanding of the options – precisely what type of fitting you are taking out, what type you are putting in, how the pipes run, how the set-up sits in relation to the room, and so on. The Expert Points opposite will describe some of the options and show you how to get things right the first time around.

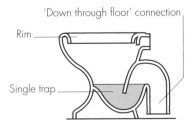

A traditional wash-down pan with 'down through floor' connection.

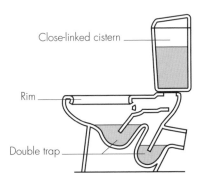

A double bend (trap) siphonic pan with close-linked cistern and 'back through wall' connection.

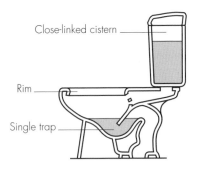

A single bend (trap) siphonic pan with close-linked cistern and 'down through floor' connection.

10 Expert Points

ALL YOU NEED TO KNOW TO FURNISH THE SMALLEST ROOM.

1 HIGH-LEVEL WC CISTERN
A high-level WC cistern is the old-fashioned type that is fixed on a bracket above head height, with a long downpipe and a chain-operated flush. This type is ideal if you are restoring a pre-1950s property. Reproduction and antique designs are readily available. Make sure that there is enough ceiling height to accommodate it.

2 LOW-LEVEL WC CISTERN
Low-level WC cisterns are the most common type. They are usually made of ceramic and fixed on brackets with a short downpipe and a lever-operated flush.

3 SLIMLINE WC CISTERN
A slimline WC cistern – a slender plastic cistern screwed directly to the wall, with a short downpipe and a lever- or knob-operated flush – is a good option if you are short of space and money. It is fragile in the sense that its structural integrity relies on the lid being correctly fitted. Slimline cisterns are not suitable for toilets that will get a lot of rough usage.

4 CLOSE-LINKED WC CISTERN
A close-linked WC cistern is designed so that the pan and the cistern are closely linked as a single feature. It is made in ceramic, with no visible downpipe, and the U-bends or traps are part of the form. It can be tricky to fit, so is not a good idea if you are a nervous plumbing novice.

5 CONCEALED WC CISTERN
Concealed WC cisterns are made in ceramic, plastic or metal, and designed to be completely hidden from view behind a panel. The downside is that the panel has to be easily removable, so that the cistern can be maintained – this is undesirable if you want to tile the bathroom.

6 WASH-DOWN WC PAN
The wash-down WC pan – a traditional design that we all recognize – has the waste pipe running either back through the wall or down through the floor. It is a good, low-cost, easy-to-fit option, but you must make sure that you get a WC pan that has a 'back through the wall' or 'down through the floor' design so that it matches the original.

7 SINGLE-BEND SIPHONIC WC PAN
The single-bend siphonic WC pan is much like the traditional wash-down WC pan, with the waste pipe running either back through the wall or down through the floor. The pan and the cistern are closely linked, and the narrow outlet at the other side of the U-bend results in a swift, sucking, siphonic action. The single-bend siphonic pan is a good choice in that it is beautifully efficient, but it can be difficult to fit.

8 DOUBLE-BEND SIPHONIC WC PAN
A double-bend siphonic WC pan is like the single-bend siphonic WC pan in all respects, but with the addition of a second U-bend. The two U-bends and the siphonic action result in a partial vacuum between the two bends. This is a super-efficient model, but it is difficult to fit.

9 WALL-HUNG WC PAN
A wall-hung WC pan is like a single-bend WC siphonic pan in most respects, except the design is such that the pan appears to be suspended from the back wall. It can be obtained as a kit, complete with metal supports and a WC cistern. It's a good choice for a modern bathroom.

10 REPEATING AN OLD DESIGN
If you want to replace an existing system with an identical unit, but are unclear about your precise needs, it's a good idea to photograph the toilet and to take the picture to a specialist supplier.

Removing a toilet

Removing a toilet from the bathroom or downstairs loo is a difficult operation, in the sense that it is messy and time-consuming. If you want the task finished in a day – and most people do – plan out the whole operation so that there are no snarl-ups. But just in case there are problems, you need to allow for the toilet to be out of action for 24 hours.

There is no clean and simple way of removing a toilet, especially if it is cemented to the floor, with the soil pipe held in place with rock-hard putty. The most important part of the process is to make sure that you don't damage the large soil outlet pipe behind the WC pan.

Method

1 Turn off the water and remove the feed pipe to the WC cistern. You may need to cap off this pipe if the pipe runs on to feed washbasins or other outlets in the house.

2 The WC cistern will be fitted to the wall with large screws that might be rusty and difficult to move. If they are rusty, you may have to rip the cistern off the wall with a crowbar, or even break it into pieces with a hammer.

3 The overflow pipe running to the outside can be simply cut off with a hacksaw or the cement can be chipped out and the pipe withdrawn from the wall.

4 Once the WC cistern is out of the way, have a good look to see how the pan is fixed to the floor. On a wooden floor it will be held down with screws, which you should be able to remove. However, it is usually quicker to simply knock it into pieces with a large hammer – wrap the pan in an old sheet (to stop the shards flying about) and wear goggles and gloves. Be very careful that you don't crack or damage the soil pipe at the back of the toilet. Use a small chisel to chip out the pieces of hard putty that line the inside face of the soil pipe.

5 If the soil pipe is of the older, ceramic type, it may break. If this happens, you will need to cut the pipe squarely to allow a modern push-fit WC connector to be used. There are two tools that can be used to cut the pipe – one is a chain-link cutter, and the other a disc cutter. Both of these can be hired from tool-hire centres. WARNING: These tools are potentially very dangerous. Ask the tool-hire centre for help, and follow their advice to the letter.

6 Push a piece of plastic sheeting into the open soil pipe to prevent bits of broken pan going in and unpleasant smells coming out.

Fitting a new toilet

It is difficult to describe how to fit a new toilet, as no two installations are the same. It is best to start by carefully measuring the diameter of the soil pipe in the wall or floor, and the distance from the outlet to the WC pan. Place the new pan on the floor in its expected position alongside the WC cistern. There are a variety of specialist connectors for joining the

WC pan to the soil pipe, which are designed to allow for small differences in height and angle.

Goes through wall – good for upstairs and downstairs toilets

Goes down through the floor – good for downstairs toilets with concrete floors

Traditional wash-down pans with single water-filled traps.

Method for siting the WC pan

1   Push the connector on to the soil pipe along with its adapter and/or bend, and ease the WC pan into place. You may need to put a little silicone grease on the connectors so that they slide on nicely.

2   If the WC cistern is of the type that sits on the WC pan itself, make sure the pan is in the right place so that the cistern can be fixed to the wall. When you're sure that the pan and cistern are going to marry up, drill holes, fit plugs, set plastic washers either side of the ceramic parts, and drive the screws into place. Do not over-tighten the screws – just tighten them enough to nip the plastic washers.

3   You may need to put the WC pan on a base of cement or silicone mastic if the floor is not level. If this is the case, draw around the pan base, remove the pan, trowel the cement in place, and install the pan as already described.

Method for connecting the pipes

1   With the water supply pipe in place (this is usually a 15 mm copper pipe), fit a tap connector on the pipe, complete with a small fibre washer to seal. This tap connector fits to the metal or plastic ballcock unit. With a new fibre washer in place, gently tighten the fitting until it is clenched securely.

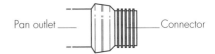

Pan outlet ____   ____ Connector

A push-fit plastic WC pan connector for a 'back through wall' connection.

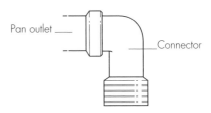

Pan outlet ____   ____ Connector

A push-fit plastic WC pan connector that converts the more common 'back through wall' pan to a 'down through floor' connection.

2   The overflow pipe (usually made of 21 mm PVC tubing) runs out through the nearest outside wall, where it protrudes by about

150 mm. If you need to fit a bend, go for the solvent-welded type.

3 When fitting the ballcock and siphon into the cistern, wipe a little silicone sealer on the WC cistern where the fitting is to go through. Clench up by hand as tightly as you possibly can.

Washbasins

Bathroom washbasins come in just about every shape, colour and size you can imagine, but your selection will be made according to a logical blend of design and function. You might like the notion of a minimal ceramic basin that appears to float out from the wall with hardly any visible means of support, but this choice will only be possible if the structure of the

bathroom wall is sufficiently strong to support a wall-hung basin.

Start by finding out about the structure of your bathroom walls. Then decide which of the three basic basin types (pedestal, wall-hung or counter-top) you want, and see if this type comes in the shape, size, form and material that you have in mind. A successful installation depends on having a clear understanding of the options – precisely which type you are taking out, which type you are putting in, how the pipes run, how the set-up sits in relation to the room, and so on. The Expert Points opposite will describe some of the options and show you how to get it right. Time spent researching all the available options is time well spent.

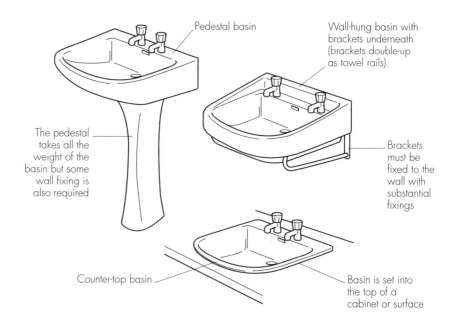

Pedestal basin

Wall-hung basin with brackets underneath (brackets double-up as towel rails).

The pedestal takes all the weight of the basin but some wall fixing is also required

Brackets must be fixed to the wall with substantial fixings

Counter-top basin

Basin is set into the top of a cabinet or surface

Fixing a new washbasin.

10 Expert Points

ALL YOU NEED TO KNOW FOR A PERFECT MARRIAGE BETWEEN BASIN AND BATHROOM.

1 BASIN TYPE
If you are replacing an existing basin, it is easiest to fit one of the same type (but not necessarily the same colour or form).

2 ASSESS THE WALLS
Have a poke around in your bathroom. Find out about the structure of the walls – especially the wall where the basin is to be placed. If you have any doubts about the solidity or condition of the walls you can fit freestanding units.

3 SOLID WALLS
Solid walls are great for wall-hung basins, where the plumbing is to be mounted on the surface, but they are a problem if you want to conceal all the pipes.But then again, you can always make a feature of the pipes.

4 HOLLOW WALLS
Hollow plasterboard walls are good for hiding pipes, but not so good if you want to have a wall-hung basin.

5 PEDESTAL BASINS
A pedestal basin is a good option in that the pedestal not only supports the weight of the basin, it also conceals all the untidiness of the pipes behind it.

6 WALL-HUNG BASINS
Wall-hung basins are a good option for small rooms where there is a shortage of floor space – because you can mount the basin so that it partially hangs over the bath. However, you must assess the walls.

7 WALL-HUNG CORNER BASINS
Miniature wall-hung corner basins are a good choice on two counts – they are ideal for very small rooms, and can be fitted both to solid walls and hollow timber and plaster walls, so are ideal for any situation.

8 COUNTER-TOP BASINS
Counter-top basins are a great idea in a large bathroom, when you want to both conceal all the pipework and have extra storage space under the basin.

9 FITTING A PEDESTAL BASIN
If you want to fit a washbasin in the shortest possible time and at the lowest possible cost, a small pedestal basin is the easiest option. The pedestal does push the price above that of a wall-hung basin of the same size, but it cuts down on the amount of work that will be involved in fixing the basin securely to the wall.

10 FITTING A COUNTER-TOP BASIN
Counter-top basins are freestanding, with their weight being taken by the structure of the cabinet, so can be fitted independently of the walls.

Fitting a new washbasin

There are three main types of washbasin.A pedestal basin is mounted on a pedestal, a wall-hung basin is mounted directly on the wall, and a counter-top basin is recessed into a slab or counter (rather like fitting a kitchen sink). Pedestal basins are good for disguising the pipework and the pedestal itself provides support for the basin, so the wall fixings do not have to be quite as substantial as for a wall-hung basin.

The main difficulty with fitting a washbasin is not so much the fitting of the feed pipes supplying hot and cold water, because these can be achieved with 15 mm copper pipe, but rather

73

how best to site the waste pipe. The problem is that the pipe carrying wastewater from the washbasin must have a gradual slope that runs towards the soil pipe (single-stack system) or large waste pipe that usually runs down on the outside wall of your house (two-pipe system). Ideally, the wastewater pipe needs a slope or drop of 6 mm for every 300 mm of pipe run, with the total length of pipe not exceeding 3 m. It may be possible to connect the sink waste to the shower or bath waste, and there are many fittings that allow you to do this.

Method for fitting a basin to a wall
1 The basin itself is usually supplied with large screws and wallplugs that allow you to fit it directly to a solid brick or block wall.
2 If you're fitting the basin to a plasterboard wall, some extra woodwork (to support the basin) will be required in the wall space. Fit some extra timber battens and a sheet of 18 mm plywood. (Screw the plywood through to the underlying structure of the wall, with screws at 150 mm centres.) Then either cover with plasterboard to give a smooth finish, or cover with tiles. Although plasterboard makes fitting that much more difficult, the advantage of it is that you might be able to conceal the plumbing in the hollow space within the wall.

Fitting other types of basin
If you have children, it is better to fit a basin with a pedestal base, as this will provide extra support and stop the basin fixings being pulled out of the wall. It is also a good idea, when you are choosing a basin, to consider how you intend to use it. Do you want to wash your hair in it? Or wash the baby? Or have a large surround for toothbrushes and suchlike? Remember to take into account the space needed around the basin – generally a width of about 1.2 m and a depth of 0.75 m, allowing you to bend over the basin comfortably.

Fitting washbasin taps
There are many different types of washbasin tap – perhaps hundreds – but the one thing that most of them have in common is a 15 mm threaded base or stem for connecting to the supply pipes. Be wary about using plastic taps and foreign imports, no matter how cheap or beautiful they may be, since they can sometimes be non-standard. Remember to check that the taps you are considering will fit the layout of the holes in the basin.

Although it's not of much concern to most people, it's interesting to know that taps work slightly differently depending on which model you buy. Old-fashioned taps (rising-spindle type) have a rubber washer inside that wears out after a time, whereas newer taps have two ceramic discs that last more or less forever. The function of an old-fashioned tap requires that you move the spindle through several turns before the water goes on or off, while a tap with ceramic discs is quicker to respond, needing only a quarter-circle turn.

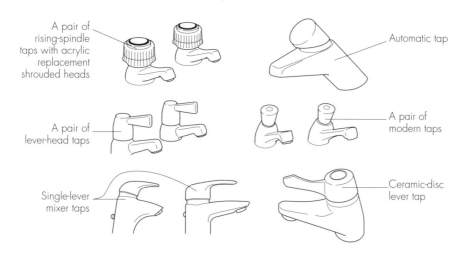

A pair of rising-spindle taps with acrylic replacement shrouded heads

Automatic tap

A pair of lever-head taps

A pair of modern taps

Single-lever mixer taps

Ceramic-disc lever tap

A range of modern basin taps.

Method

1 The tap fits through from the top of the basin, usually with a small rubber seal between the body of the tap and the basin. The washer and large nut that are usually supplied with the tap are screwed into place on the underside of the basin. You will need a basin spanner (cranked spanner) for this operation – these are cheap to buy.

2 If you're replacing an existing basin, simply reconnect the existing feed pipes for hot and cold water to the bottom of the new taps. Make sure you put a new fibre washer on each tap (these look like they are made of red cardboard). The pipe back-nuts need to be tightened firmly, but not until they will not budge at all!

3 If you're fitting a new basin, bring the plumbing to within 200 mm of the bottom of the taps and then fit a flexible connector. This spacing allows for a small amount of misalignment between the pipe and the taps. There are two types of flexible connector. One is a corrugated copper unit that is easily bent into shape and then soldered on to the pipework; the other is in the form of a flexible rubber pipe with braided stainless-steel wire sheathing. Such connectors have either compression or collet-type push-fit fittings, and can be purchased with or without an integral service valve. (Service valves are particularly useful in that they allow you to service the tap simply by shutting off its own dedicated valve with a coin or screwdriver. This is a good option in an emergency.) Flexible connectors are expensive, but are most definitely worth the extra cost.

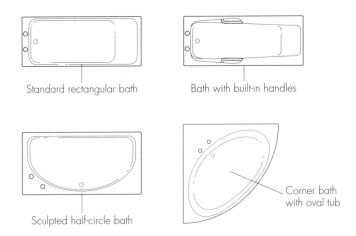

Standard rectangular bath

Bath with built-in handles

Sculpted half-circle bath

Corner bath
with oval tub

Acrylic baths can be obtained in many different colours, designs, sizes and styles.

All about baths

Replacing a bath, or perhaps fitting out a second bathroom, is one of the biggest plumbing statements that you can make – it's one plumbing job that is really noticed and appreciated. The actual procedures are pretty straightforward. The tricky part is working out how to get the old bath out and the new one in. However, if you spend a lot of time researching, measuring and planning, it's actually very easy. The Expert Points opposite will help you get it right.

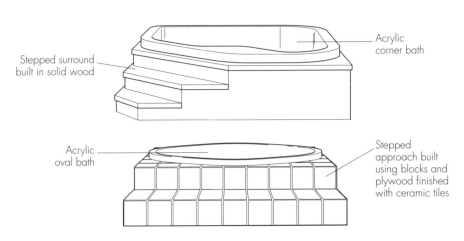

Acrylic
corner bath

Stepped surround
built in solid wood

Acrylic
oval bath

Stepped
approach built
using blocks and
plywood finished
with ceramic tiles

Acrylic baths set in raised, stepped surrounds.

10 Expert Points

CONSIDER THESE POINTS BEFORE
MAKING A FINAL DECISION ABOUT THE
TYPE OF BATH THAT YOU WANT.

1 CAST-IRON BATH
A traditional cast-iron bath with a
white vitreous enamel finish is, in our
opinion, the very best option. Once a cast-
iron bath is in place, it is pretty immovable.
This makes life difficult if you are thinking of
taking out an old one or putting in a new
one. You will need three or even four people
to heave and drag it into place.

2 BREAKING UP A CAST-IRON BATH
If you want to get rid of an existing
cast-iron bath, and you are not intending to
sell it, you can break it up with a club
hammer. (Protect your eyes with goggles,
and cover the bath with a dampened old
blanket. WARNING: The blanket and
goggles are essential for your protection,
since they will protect you from flying
shards of razor-sharp enamel.)

3 SELLING A CAST-IRON BATH
You might hate the sight of your cast-
iron bath, but there is probably someone
restoring an old house who wants it. Old
cast-iron baths in good condition, with no
cracks, can be worth money.

4 RENOVATING A CAST-IRON BATH
Old cast-iron baths with damaged
enamel can be renovated *in situ* by a
specialist firm. They will clean the surface,
spray on a new enamel-like acrylic finish in
the colour of your choice, fit new taps and
so on. This might be a good option.

**5 STRUCTURAL REQUIREMENTS
FOR A CAST-IRON BATH**
If you want to fit a new cast-iron bath in an
upstairs room of an old house, you must
check the strength of the floor. Get a builder
to check it if you are unsure.

6 ENAMELLED PRESSED-STEEL BATH
An enamelled pressed-steel bath –
very much like a cast-iron bath in shape and
character – is a good choice if you want to
achieve the effect of cast iron without the
high cost and handling difficulties. Two
people can move a steel bath easily.

7 PLASTIC BATH
A plastic bath – made of acrylic or
plastic and reinforced with glassfibre – is a
good, low-cost, easy-to-handle choice if you
want a bath in an exotic shape or colour.
Two people can move a plastic bath with
ease, even if it's a giant size.

8 CAST-IRON ROLL-TOP BATH
Cast-iron roll-top baths with a rounded
end were designed to be freestanding, with
the underside and feet on view. Be warned:
if you plan to box in a bath with a rounded
end, it is a difficult task.

9 MEASURING THE SPACE
Always measure your bathroom very
carefully before you get a new bath.

10 PANELLING A BATH
If you are planning to fit end and side
panels to the bath, make sure you get
panels that will suit the overall design and
colour of your bathroom.

Removing a bath

The process of removing a bath is
usually straightforward, but because it
is large and cumbersome you are
likely to need someone to help you.

Method

1 If there are small service valves
on the supply pipes to the bath,
simply turn these off with a coin
or screwdriver and disconnect the

taps, otherwise you'll need to turn off the water supply and drain the system by running a sink or hose at the lowest point in the pipes.

2   If the new bath is going in exactly the same place as the old one, you may be able to leave the existing supply pipework in place. If this isn't the case, cut it off about 150 mm below the taps. The waste pipe will certainly be wrong, so simply cut it off with a hacksaw. Keep a washing-up bowl ready to catch any water, and have plenty of towels to mop up with. Some older baths have an overflow pipe that runs directly from the overflow through the outside wall. You will not need this with a modern bath, so simply pull it out of the wall and fill the hole with expanding foam or cement. You may need to cut the waste pipe or supply pipes back further, but it is best to sit the new bath in place first so that you can see how it relates to the existing pipes, and then connect the new and existing pipes to suit the existing ones.

3   Baths can be quite difficult to remove from the wall, due to hidden brackets, tiling cement, mastic and so on. Look under the bath to see whether it has small, adjustable feet. If it has, screw these up so that the bath drops downwards. Stand in the bath so that your weight cracks the bath away from the tiles and mastic. Being very careful not to catch or move the pipes, slide the bath away from the wall.

A typical pre-1940s bathroom with cast-iron roll-top bath and heated towel rail.

A typical 1950s bathroom with enclosed cast iron bath, washbasin on a stand and a low-level close-linked WC cistern.

4   If the discarded bath is made of cast iron and you plan to throw it away rather than sell it, it is usually easier to smash it up in order to take it out of the house (see Expert Point 2 on page 77). The broken pieces will be very sharp and heavy, so take great care when moving them.

5   If you are removing the old tiles from around the bath, now is the time to do it. Make all the mess and repairs to the plaster before

the new bath is put into position, otherwise you risk damaging it by dropping a heavy tool or old tile into the bath while working.

Fitting a new bath

Whether you're fitting a new style of bath or replacing an old one with an identical version, the method of working is the same.

The water supply pipes to the bath should always be in 22 mm copper, so that there is a decent flow of water from the taps. This may mean that you have to install new pipes right back to the hot-water cylinder or the cold-

water tank in the loft, depending on which system you have in your home. If your existing bath already has 22 mm pipes, you will need to fit flexible tap connectors of a suitable length to reach the new taps (which may be in a slightly different position). The different types of flexible tap connector are described in 'Fitting a washbasin' on page 75.

Modern baths usually have adjustable feet, so that you can alter the height of the bath to allow for an uneven floor. If the bath is to stand on a timber or chipboard floor, you will need to place a couple of pieces of

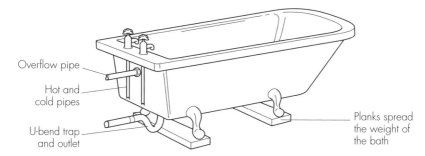

Overflow pipe

Hot and cold pipes

U-bend trap and outlet

Planks spread the weight of the bath

Typical connections to a cast-iron bath.

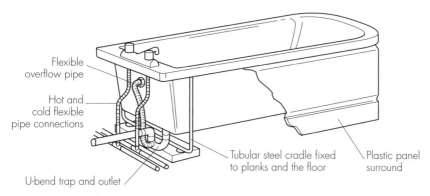

Flexible overflow pipe

Hot and cold flexible pipe connections

U-bend trap and outlet

Tubular steel cradle fixed to planks and the floor

Plastic panel surround

Modern plastic bath.

18-mm-thick plywood underneath the bath feet in order to spread the load over the joists. Sit the bath in position, place a spirit level on the rim of the bath, and adjust the feet so that the bath is level in its length and width. Don't put the spirit level in the bottom of the bath as this has a built-in slope to allow for drainage; always take the levels from the rim!

Method

1 Buy a kit for the 40 mm waste outlet in the bottom of the bath. Fit it to the bath, making sure you have the seals in the right place, and applying a smear of silicone sealant over the washer faces. Make sure that you don't over-tighten the nut on the waste outlet, especially if the fitting is made of plastic. Some waste outlets have a screw that is put in from the bath side, which clamps the waste outlet fitting together. If there are instructions and a diagram on the packaging, study them carefully before starting work.

2 Most kits come with an overflow fitting and pipe to take overflow water down to the waste outlet fitting under the bath. Fit the waste trap and wastewater pipe (40 mm diameter), and connect up to the soil pipe (single-stack system) or large waste pipe (two-pipe system), just as you would a washbasin.

3 Run water into the bath and check underneath for leaks. When you pull out the plug and let the water drain away, it should drain away quickly, with no sign of water

under the body of the bath. If you do notice a slight leak, the likelihood is that it will be in the connection of the waste trap or possibly where the overflow pipe connects under the bath.

4 When you're happy that all the pipework is operating properly and without leaks, you can fit panelling around the bath. Some bathroom suites come complete with side panels made of thin plastic. If you want to fit such plastic panels, you need to build a support framework under the bath. The framework should be made from 50 mm x 25 mm pine. Start by screwing lengths to the floor and walls, using a spirit level to make sure everything is upright and square. Measure and cut the vertical sections, and connect these to the other parts of the framework using screws and/or small metal strips and corners. Ideally, you need a framework around the open sides of the bath that goes from the floor to just under the rim, with several uprights spaced along the bath length, and maybe one on the end. It doesn't have to be beautiful, as long as everything is firm, straight and upright. Fix the bath panels with small brass screws, so that they can be easily removed for servicing and maintenance. The panel at the end of the bath can be permanently fixed, as long as the tap and water trap end of the bath is accessible through the side panel. You could fit small cupboards under the bath.

All about showers

A shower is very efficient in terms of water usage, whereas a full bath contains enough water for about four showers. A shower also takes up very little room, so is also suitable for fitting in the corner of a large bedroom. There are three common types: the gravity-fed shower, the power shower and the instantaneous electric shower.

Shower types

Gravity-fed showers use stored water that is not at mains pressure. The cold feed comes from a header tank in the loft (the cold-water storage tank), and the hot feed comes from the hot-water cylinder or combination boiler. For this type of system to work effectively, the header tank needs to be at least 1 m above the showerhead in order to provide sufficient pressure and a reasonable flow of water.

A power shower is the same as a gravity-fed shower, but with a hidden electric pump that boosts the pressure to the showerhead to give a more invigorating spray.

Some types of shower use water directly from the cold-water mains, which is at mains pressure. A branch pipe from the rising main takes it to the shower. Instantaneous showers and thermal-store showers work in this way. This type of shower can be fitted in the topmost room of the house, regardless of the position of the header tank, as long as there is plenty of headroom.

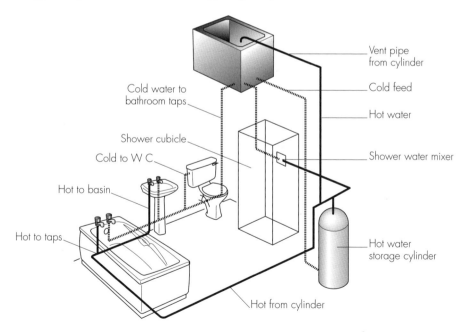

Typical arrangement for a gravity-fed shower.

Gravity-fed shower system layout

In most houses, a gravity-fed shower system takes hot and cold water at low pressure from the same supply that goes to the bath taps. Generally, both the hot- and cold-water supply is put under pressure by the weight of water in the loft header tank. The water flows to a mixer tap or thermostatic mixer where the temperature is adjusted for comfort. From the mixer tap, the warm water flows through a single pipe to the showerhead. The spray from this type of installation may be relatively weak without a booster pump to increase the pressure.

Instantaneous electric shower system layout

Typical electric instantaneous shower unit.

An instantaneous shower takes mains water (at high pressure) through a single 15 mm pipe into a small copper cylinder in a wall-mounted unit. When the shower is turned on, a pressure sensor in the unit activates an electric heater in the cylinder. The water temperature is adjusted by regulating the amount of cold water passing to the showerhead through the heater. The disadvantage of cheap instantaneous showers is that during a really hot shower less cold water is needed, so its flow is slowed down with the result that the spray is weak. However, these showers are inexpensive and quick to install.

Showerheads

Not long ago, there was a very limited range of showerheads available, but it is now possible to get them in many shapes and sizes: some are no more than a plate with holes, while others can be adjusted to govern the shape, form and intensity of the water spray. The children might enjoy a light mist of water, but you might prefer a concentrated jet.

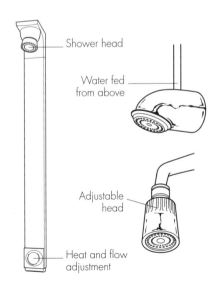

Shower head

Water fed from above

Adjustable head

Heat and flow adjustment

Modern shower head fittings.

10 Expert Points

THE FOLLOWING TEN POINTS WILL HELP YOU INSTALL A SHOWER.

1 WASTE PIPE
A shower requires a 40 mm waste pipe, so make sure that there is a sensible route for this pipe from the shower to the nearest soil pipe (single-stack system) or large waste pipe (two-pipe system).

2 PROVIDING HOT WATER
Decide whether you want an instantaneous shower, which uses electricity to heat water, or a shower that uses your existing hot-water supply.

3 SITING THE SHOWER
Shower trays are very noisy when the water is running. If a bedroom is next door, the shower might keep the occupant awake.

4 DEALING WITH A CONCRETE FLOOR
If the floor is concrete, the shower tray will have to be raised on a plinth (10 cm high), or a channel will have to be cut in the floor to provide clearance for the drain.

5 POSITIONING A PUMP
In the best power showers, the electric pump is hidden. The pump is about the size of a shoebox and needs to be concealed in the wall, ceiling or a cupboard.

6 HIDING PIPEWORK
If there is nowhere else to hide the pipes, a small partition wall can be built (10 cm away from the existing wall) to make a channel in which to run the pipes. This type of partition can be made easily with a basic 50 x 100 mm studwork frame and plasterboard or, better still, a covering of plywood. The partition should be tiled to ensure the area is fully waterproof once the shower is fully fitted and tested.

7 BEDROOM SHOWER
A plastic or safety-glass cubicle allows a shower to fit neatly into a bedroom. Wardrobe-style doors can be easily fitted around this to disguise the entire installation.

8 ELECTRICITY SUPPLY
This book will show you how to fit a shower that uses electricity, and how to do most of the necessary wiring, but you should not make the final connection to the consumer unit (fuse box). A professional electrician needs to make final connections and test the wiring.

9 SHOWER ATTACHMENTS
Many bath taps can be replaced with a simple combination bath mixer tap and shower. This means that you can have a shower for the cost of a set of new taps. The new taps will need to operate at a similar pressure to the ones you are replacing, and the showerhead will need to be at least one metre below the header tank. See 'Fitting a power shower' on page 87.

10 INSTALLATION COST
The cheapest complete freestanding shower would be to tile two sides of the corner of a room, fit a curved or angled curtain rail to the wall for a plastic curtain for the other two sides of the 'cubicle', and fit an instantaneous electric shower unit. The cheapest shower trays are made of fibreglass, which has a warm feel.

The performance of the shower is, to a great extent, governed by the condition of the showerhead, so it is vital that the head is cleaned at regular intervals, particularly in areas where there is hard water. Showerheads are either held together with a central screw, made up of pierced plates or discs that screw together, or simply come in the form of a fixed rose.

Method for cleaning a showerhead

1 Unscrew the whole head from the shower and take it to the kitchen sink. Take the unit apart, and then clean the various components with a non-abrasive brush (such as an old toothbrush).
2 Put the components in either vinegar or a proprietary descaler, and leave to soak overnight. (Read the instructions on the descaler carefully, to ensure that it will not harm your showerhead – some materials will be affected detrimentally by certain substances and chemicals.)
3 Rinse the components under running water, put them back together and fit the showerhead back on the shower.

Shower mixers

In a shower, the water flows to a mixer tap or thermostatic mixer where the temperature is adjusted for comfort. From the mixer, the warm water flows through a single pipe to the showerhead. There are three types of mixer: a combination bath mixer tap and shower, a manual shower mixer, and a thermostatic mixer.

Combination bath mixer tap and shower

The temperature of the water is adjusted by turning the hot and cold taps until the water running from the mixer tap is a suitable temperature, and then operating a button to divert the water to the showerhead. This is a good option if you want to keep the running costs down.

Manual shower mixer

A manual mixer features a single control for regulating the temperature and flow of water in the shower. While this is a good low-cost option, it's not such a good idea if the shower is intended to be used by children or elderly people.

Thermostatic shower mixer

Thermostatic shower mixers incorporate a control to preset the temperature of the water. They are a little more expensive than ordinary mixers, but are well worth the extra financial outlay on many counts. They are much safer to use than other types of shower because the temperature is regulated. If someone uses hot or cold water elsewhere in the house, a thermostatic mixer instantly adjusts the water flow so that the temperature of the shower remains the same. The temperature dial operates independently of the on/off control, so the temperature setting can be maintained when the water is turned off. When you next use the shower, the temperature will be exactly the same. This is a good option when it is intended that the shower will be used by children or elderly people. It is also possible to get fail-safe units that can be locked to a safe temperature.

The mechanism inside a thermostatic mixer shower is complicated. The temperature is controlled by a small wax field piston or a bi-metallic strip mechanism, which continually adjusts the flows from the hot and cold pipes to give a constant outlet temperature.

10 Expert Points

THERMOSTATIC SHOWER MIXERS MAY SOUND COMPLICATED. THESE POINTS WILL HELP CLEAR UP ANY CONFUSION.

1 SHOWER TYPES
Thermostatic mixer units work with either gravity-fed showers or pump-assisted power showers. Before buying the mixer unit, check the manufacturer's specifications and make sure that there is enough height between the shower and the cold-water storage tank in the loft.

2 BOOSTER PUMP
If there is not enough height for a gravity-fed system, you can install a normal twin-impeller booster pump that turns itself on when the shower is turned on. Refer to 'Fitting a power shower' on page 87.

3 TEMPERATURE TROUBLESHOOTING
If your thermostatic mixer shower seems to be either too hot or too cold, it may be that the hot-water temperature going into the shower is too high and the mechanism cannot adjust sufficiently. Set your hot-water temperature in the boiler or immersion heater to about 50°C. This should cure the problem.

4 PRESSURE PROBLEMS
Thermostatic mixer showers do not work well when there is a big difference between the pressure of the hot and cold water.If you think that this could be a potential problem, remember that many shower manufacturers are only too willing to send out a representative to advise you on the best option for your situation.

5 POOR WATER FLOW
If you have recently installed a thermostatic mixer and you find that the water pressure from the showerhead is very weak, install a twin-impeller booster pump on the hot- and cold-water feed pipes to increase the shower pressure.

6 LIMESCALE BUILD-UP
In a brand new thermostatic mixer, the mechanism will respond very quickly. However, over a period of time, the build-up of limescale will slow the movement of the mechanism, and you will feel an occasional variation in temperature. If this happens, overhaul the unit.

7 CLEANING THE MIXER UNIT
If the mixer unit is becoming ineffective due to limescale build-up, soak the component parts in a weak solution of kettle de-scaler and use an old toothbrush to clean away the debris. Read the instructions carefully beforehand, as the solution may react badly with some metals. Wear goggles and gloves.

8 BRANDS
If possible, buy a top-quality, brand-name thermostatic mixer shower, so that in the event that you need spare parts three or four years down the line, they are more likely to be readily available.

9 SERVICE VALVE
When you fit a thermostatic mixer shower, install a small service valve in both the hot- and cold-water feed pipes. The addition of these valves will allow you to service the shower without having to turn off the water supply to the whole house.

10 LOCATING THE SHOWER
Ideally, the thermostatic mixer shower needs to be connected midway between the hot-water cylinder and the cold-water storage tank in the loft. This location will give the best possible flow to the shower unit, and consequently the best temperature-control performance. Before deciding where to put your shower, look at all the available technical literature that relates to the unit you have chosen. If you still have any doubts about where to locate the shower, ask a plumber for his advice.

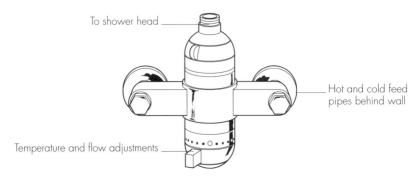

To shower head

Hot and cold feed pipes behind wall

Temperature and flow adjustments

Thermostatic shower fitting (Option 1).

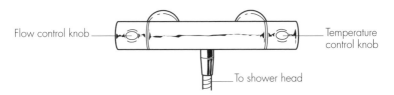

Flow control knob

Temperature control knob

To shower head

Modern thermostatic shower fitting (Option 2).

Thermostatic showers: options and types

The mechanism in a thermostatic shower mixer does not vary much between different makes; however, there are many options when it comes to controls, pipework connections and material specifications.

Option 1

With this unit, as illustrated, the pipework is hidden in the wall. The unit has two rotary dial controls: one to set the temperature, the other to control the flow of water from 'off' to 'maximum'. This is a good option when the wall is a hollow partition wall, or where it is made of plasterboard. If you want to fit a unit of this type on solid walls, you will have no choice but to channel the wall and make good with plaster.

Option 2

With this unit, as illustrated, just about everything is chrome-plated – the feed pipes, showerhead and body of the mixer unit. This type of installation is suitable for solid or partition walls. Note that the hot feed pipe will get very hot and should not be touched – so perhaps this type should be avoided if you have young children. The controls for this sort of unit may also be either in the form of a rotary dial (as shown in Option 1) or an all-in-one joystick. With a joystick, the stick is lifted to turn the water on, and swung left and right to change the temperature. The chrome finish on this mixer unit will emphasize limescale and soap deposits, so it will need regular cleaning to maintain the high-shine finish.

Fitting a power shower

● Ideally, the shower needs to be sited as near as possible to both the hot-water cylinder and the cold-water supply tank in the loft, so that variations in pressure (when other toilets or taps are used) are kept to a minimum. Be sure to read the literature supplied with your unit.

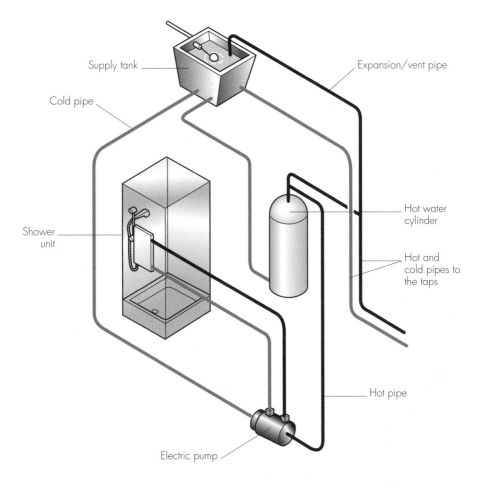

Supply tank

Cold pipe

Expansion/vent pipe

Shower unit

Hot water cylinder

Hot and cold pipes to the taps

Hot pipe

Electric pump

Layout for power shower.

- Always fit stopcocks or service valves near to the cylinder and tank, so that the shower unit can be serviced easily when necessary.
- As a power shower pump creates suction in the supply pipes, you may have to fit a new, dedicated connection a little way down from the top of the cylinder – a Surrey flange. This unit will prevent the power shower from sucking water from the cylinder vent.
- The electrical supply to the pump needs to be connected to a double-pole ceiling switch (an extra-safe switch, with a mechanism that breaks both the live and neutral conductors in the circuit) and a fused connection unit (FCU). There needs to be a residual current detector (RCD) in line in the supply wiring between the FCU and the consumer unit, so that you will be protected from any electrical faults.

WARNING: The electrical work must be done by a qualified electrician, unless you are absolutely competent in this area of expertise. Read the manufacturer's instructions carefully and make sure the shower pump unit is fitted properly.

Method for connecting to the hot-water cylinder
1 Turn off the water supply to the hot-water cylinder and open the cylinder's draincock to lower the water level sufficiently.
2 Unscrew the vent pipe at the very top of the hot-water cylinder. Take a Surrey flange, wrap PTFE tape

around the threads, and screw it into the top of the tank. Screw the vent pipe securely into the top of the Surrey flange.
3 Fit the supply pipe to the shower to the side of the hot-water cylinder, with a small stopcock or service valve fixed in line (no further than 1 m away from the hot-water cylinder).

Method for connecting to the cold-water tank
1 Turn on the cold-water tap in the bathroom, and turn off the tap in the pipe that runs to the cold-water storage tank in the loft. Wait for the water tank to empty.
2 Using a hole-cutting tool, cut into the side of the cold-water tank at a point about 100 mm from the bottom. Fit a tank fitting.
3 From this tank fitting, first install a small stopcock or service valve, and then continue the pipework on to the shower unit. Tapping into the tank as described (rather than running a branch pipe from one or other of the 22 mm pipes that go into the bottom of the cold-water storage tank), means that the water supply should be free from pressure variations.

Shower design
A shower area is no more than an enclosed space with waterproof walls and a base to collect and direct the wastewater down a waste pipe, so it can be just about any shape and design that you can imagine. At one time you were limited to having a

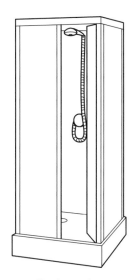

Easy-to-install cabinet shower unit kit.
Suitable for bedroom installation.

A modern shower
fitting with concealed
pipework

Old-fashioned cast-iron bath with
shower over. Requires shower curtain.

cubicle or a plastic curtain hanging in
the bath, but now you can have
anything from a dedicated room with
tiled walls and a waste trap set in the
centre of the floor, to a purpose-built
kit that can be positioned as a freestanding unit in the middle of the
bedroom floor. However, your choice
will be governed by the size of your
home, the amount you want to spend,
and how much you like taking a
shower. Bear in mind that many
options specify the ideal location and
environment – upstairs, concrete
floor, tiled walls etc.

With so many things to consider,
the Expert Points overleaf will help
you decide which shower is for you.

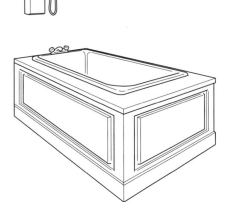

Modern plastic bath with instant electric
shower. Requires shower curtain.

10 Expert Points

CONSIDER THE FOLLOWING POINTS BEFORE YOU SETTLE ON A DESIGN FOR THE SHOWER INSTALLATION.

1 SHOWER ROOM

If you have plenty of space, and are not keen on having a shower with a plastic curtain that invariably tries to stick to your body, one of the most exciting options is to build a shower room. This will undoubtedly involve building block walls, setting a waste trap in place under a raised floor, and tiling walls and floors, but is a relatively low-cost option – far less expensive in materials than buying a large, purpose-built shower unit.

2 SHOWER IN THE BATH

If you are short of both space and cash, fit a combination bath mixer tap and shower in place of bath taps, and fit a plastic curtain that will hang inside the bath. The downside of this design is that it's not easy to stand in a curve-bottomed bath tub, and the shower curtain will inevitably tend to flap and get in the way.

3 BEDROOM SHOWER UNIT

The advantage of a unit built or set into the corner of a large bedroom is that two walls are already in place. All you do is construct a third wall running out from one of the existing walls, and then fit a curtain or door across the entrance.

4 FACTORY-BUILT SHOWER UNIT

Factory-built shower units are very expensive, but their advantage is that they can be fitted and in service within the space of a weekend. Do lots of research before choosing a unit.

5 SHOWER TRAY

A great part of the integrity of a shower unit hinges on the design of the shower tray, so it pays to get the best one that you can afford to buy.

6 PLASTIC SHOWER TRAY

Plastic trays usually score highly in terms of the availability of exciting colours and designs, and the fact that they can easily be moved and transported. They are also very warm underfoot. The downside is that, once installed, they frequently move and creep, making it difficult to create an effective seal around the edges.

7 METAL OR CERAMIC SHOWER TRAYS

Metal or ceramic trays tend to be conservative in design, expensive and difficult to transport, and cold underfoot. However, when they are in place, they stay put and are easy to seal.

8 LOCATION

The location of the shower will, to some extent, govern its structure and form. For example, it is very important that an upstairs shower is totally waterproof – more important than if the shower was sited downstairs on a concrete floor. Be mindful of this when you are thinking about siting and materials. That said, there are now some wonderful kits and units in the market - perfect for just about every location. Time spent researching all the available possibilities will be time well spent.

9 SAFETY IN THE SHOWER

If the shower is going to be used by elderly people, you ought to provide grab bars. These bars or handles need to be plentiful, strong, and securely fitted at about chest height. A non-slip surface should be applied to the shower tray to reduce the risk of accidental falls.

10 HEATING AND VENTILATION

Shower areas need to be well heated and well ventilated, with a window or extractor fan. Many units and kits come complete with heating and ventilation options – perfect for the DIY plumber.

10 Expert Points

DISCOVER ALL YOU NEED TO KNOW ABOUT THE MYSTERIES OF THE BIDET.

1 TYPES OF BIDET
There are two types of bidet: the over-rim bidet, which is just like a low-level washbasin with hot and cold taps, a plug and an overflow, and the under-rim bidet, which is fitted with mixer taps, a spray and a remotely-operated plug.

2 OVER-RIM BIDET
This simple washbasin-type bidet is low cost and easy to fit; the downside is that the top rim always feels cold to the touch each time you sit on it.

3 UNDER-RIM BIDET
This type of bidet is expensive and more difficult to fit, but the advantage of its design is that warm water runs under the rim and heats it, so that it feels warm to sit on.

4 PIPEWORK
An over-rim bidet can take its cold water from the nearest convenient pipe, but an under-rim bidet must take its cold water from a dedicated pipe that links directly to the cold-water storage tank in the loft.

5 HEALTH AND SAFETY
There are two health and safety concerns regarding small children, who might think a bidet makes a fun plaything. First, the water might come out too hot, and

second, while the height of a bidet makes it perfect for water play, its function makes it totally unsuitable.

6 WATER AUTHORITY REGULATIONS
Under-rim bidets are often described as unsuitable for DIY installation, but this doesn't mean that you are not allowed to fit one. It is because water authority regulations require that the cold water comes directly from the cold-water storage tank in the loft (and no other connections to this supply pipe are permitted), and this involves a lot of extra work.

7 POSITION
Just like WC pans, you can have a floor-standing or wall-hung bidet.

8 SURROUNDS
You should tile around the bidet, and install a towel rail at a convenient level.

9 COST OF AN UNDER-RIM BIDET
Remember that an under-rim bidet is not only a much more expensive item in its own right, but also that the extra pipes and the mixer tap are all going to add significantly to the expense.

10 PLUMBING REGULATIONS
If you have any doubts about the rules and regulations regarding fitting a bidet, it's a good idea to make contact with your water supply company.

Bidets

The dictionary somewhat mystifyingly describes a bidet as being 'a vessel for bathing in, on a low stool that can be bestridden', but granny's description of 'a basin for washing your rude bits' is much more to the point. A bidet is a bit like a WC pan that's been fitted with hot and cold taps and, if you are

lucky, an upward-pointing spray. And if you do a lot of walking or just suffer from hot feet, a bidet doubles up very nicely as a footbath!

A bidet is still considered to be something of a luxury or even a bit 'continental', but is without doubt a really useful extra. The Expert Points above will demystify the bidet.

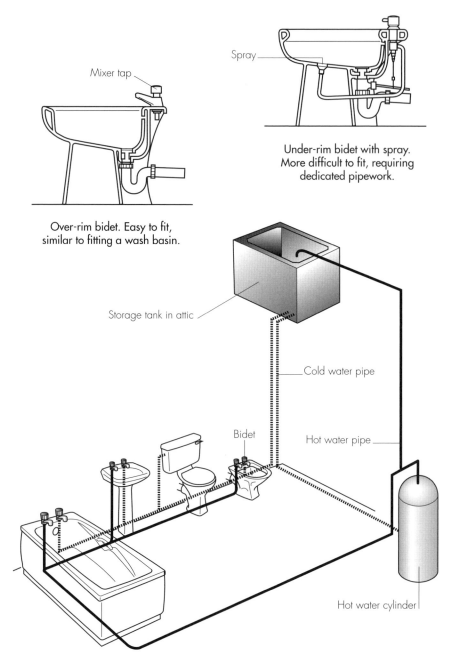

Mixer tap

Spray

Under-rim bidet with spray.
More difficult to fit, requiring
dedicated pipework.

Over-rim bidet. Easy to fit,
similar to fitting a wash basin.

Storage tank in attic

Cold water pipe

Bidet

Hot water pipe

Hot water cylinder

Fitting an over-rim bidet.

Over-rim bidet

This type of bidet is, in form and function, very much like a washbasin or bath – there are two taps and a waste outlet with a plug. The supplies of hot and cold water can be taken from the nearest source – for example the 15 mm pipes for hot and cold water that run to the bathroom washbasin. To use the bidet, you adjust the flow of water so the temperature suits your needs.

Under-rim bidet

Because this type of bidet has a sprayhead that can be submerged, with the possibility that contaminated water could get drawn back into the system, water authority regulations require that the cold-water delivery pipe comes directly from the cold-water storage tank in the loft (so the pipe is dedicated to the bidet). To use the bidet, you set the mixer tap to a comfortable temperature.

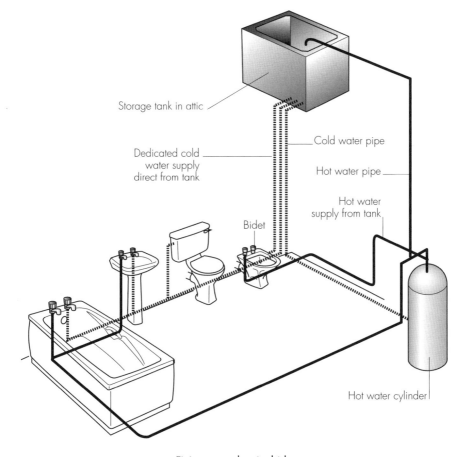

Fitting an under-rim bidet.

Fitting a new over-rim bidet

The main difficulty with installing an over-rim bidet is how best to site the waste pipe. The problem is that the pipe carrying the wastewater must have a gradual slope that runs towards the soil pipe (single-stack system) or large waste pipe that usually runs down on the outside wall of your house (two-pipe system). Ideally, the pipe needs a slope or drop of 6 mm for every 300 mm of pipe run, with the total length of pipe not exceeding 3 m. It may be that you can connect the bidet waste to the shower or bath waste – there are many fittings that allow you to do this.

Method

1 Allow sufficient space around the bidet. You generally need a width of 1.2 m and a depth of 0.75 m. This allows you to sit and stand without banging knees or ankles.

2 The bidet is usually supplied with large screws that allow you to fit it directly to a wooden floor. If you're fitting it to a solid concrete floor, use screws and wallplugs.

3 There are many different types of bidet tap – perhaps hundreds – all of which have a 15 mm threaded joint at the base for connecting to the supply pipes. Taps work slightly differently depending on which model you buy. Old-fashioned taps (rising-spindle type) have a rubber washer inside that wears out after a time, whereas newer taps have two ceramic discs that last more or less forever. The function of an old-fashioned tap requires that you move the spindle through several turns before the water goes on or off, while a tap with ceramic discs only needs a quarter-circle turn.

4 The tap fits through the top back edge of the bidet, usually with a small rubber seal between the body of the tap and the body of the bidet. The washer and large nut (usually supplied with the taps) are screwed into place on the underside of the basin. You will need a basin spanner (cranked spanner) for this operation.

5 If you're replacing an existing bidet, reconnect the existing hot and cold pipes to the bottom of the new taps. Make sure you put a new fibre washer on each tap (these look as if they are made of red cardboard). The pipe cap-nuts should be tightened firmly, not until they will not budge at all!

6 If you're fitting a new bidet, bring the plumbing to within 200 mm of the bottom of the taps and then fit a flexible connector. This spacing allows for a small amount of misalignment between the pipe and the taps. There are two types of flexible connector. One is a corrugated copper unit that is easily bent into shape and then soldered on to the pipework; the other is in the form of a flexible rubber pipe with braided stainless-steel wire sheathing. Such connectors have either compression or collet-type push-fit fittings, and can be purchased with or without an integral service valve. (Service valves are particularly useful in that they

allow you to service the tap simply by shutting off its own dedicated valve with a coin or screwdriver. This is handy in an emergency.) Flexible connectors are expensive, but are most definitely worth it.

Kitchen sinks

The kitchen sink comes with lots of symbolic baggage – 'tied to the kitchen sink', 'kitchen sink drama' – that describes attitudes to life. And most people have very strong views when it comes to the kitchen sink – the type, material, colour, size, place in the kitchen, type of taps, number of bowls and draining boards and so on. So, if you are going to fit a kitchen sink, you have got to get it right in every respect. The Expert Points overleaf will show you the best way through the sink minefield.

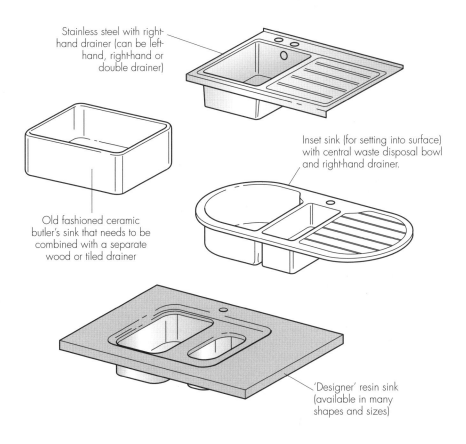

Stainless steel with right-hand drainer (can be left-hand, right-hand or double drainer)

Inset sink (for setting into surface) with central waste disposal bowl and right-hand drainer.

Old fashioned ceramic butler's sink that needs to be combined with a separate wood or tiled drainer

'Designer' resin sink (available in many shapes and sizes)

Kitchen sink types.

10 Expert Points

WE ALL SPEND HOURS AT THE KITCHEN SINK – SO MAKE IT A FITTING YOU LIKE THE LOOK OF!

1 BUTLER'S SINK

Ironically, the sink that previous generations would have thought of as being the very epitome of drudgery and general grease, grunge and grind – the glazed, white, fireclay butler's sink – is now considered to be a must-have item. Such sinks can be obtained new, or from architectural reclamation companies.

2 COST OF A GLAZED SINK

If you plan to build a country cottage kitchen complete with an old-fashioned glazed sink, wooden draining board and brass taps, bear in mind that such sinks are so heavy that they need to be supported on a sturdy wooden frame. This set-up will be relatively low in cost if you are prepared to search out a reclaimed sink and do the woodwork yourself; otherwise it will be at the top end of the price range.

3 WASTE OUTLET FOR GLAZED SINK

As old glazed sinks have very large waste holes, you will have to search quite hard to find a waste outlet to fit. If possible, see if you can find an old fitting in an architectural salvage outlet.

4 STAND-ALONE SINK AND DOUBLE DRAINING BOARD

If you enjoy cooking, you cannot do better than get a stand-alone stainless-steel sink with a single bowl at the centre and a draining board at each side. This design suits both left-handed and right-handed washer-uppers – stack the dirty crocks on one draining board, wash them in the bowl, and then stack the clean pots on the other board. Stainless steel is one of the more practical materials for kitchen sinks – it's easy to clean and doesn't stain or crack.

5 SINK WITH SINGLE DRAINING BOARD

A single-bowl, stainless-steel sink with a drainer to one side may appear to be a good choice for a minute kitchen, but has drawbacks. There is nowhere to put the dirty crocks, and if the potential washer-uppers are not either right-handed or left-handed to match the sink, it will be inconvenient.

6 SURFACE SINKS

Surface sinks (metal or resin sink set flush into a continuous surface) look good, but the draining boards tend to be so shallow that they dribble water on to the user and on to the floor.

7 TAPS

When choosing taps, remember that the flow of water needs to be at about the centre of the sink. To avoid splashing, choose a tap with a spout length to suit the design of your chosen sink.

8 TAP STYLES

There are hundreds of different taps on the market, but remember that ultra-modern taps are likely to date very quickly. The traditional capstan-head tap, on the other hand, is now so old a design that it is considered to be beyond fashion, and has become a classic.

9 TAPS FOR TWO-BOWL SINKS

If you would like a stainless-steel sink with two bowls, and you want the option of running water in either bowl, you must fit a mixer tap with a swivel spout that will move smoothly between the two bowls.

10 FITTING TAPS

Some stainless-steel sinks have pre-cut holes for the taps, while others come with a tool that allows you to position holes to suit your needs (a good option, but it is not easy to cut the holes and the sinks are expensive).

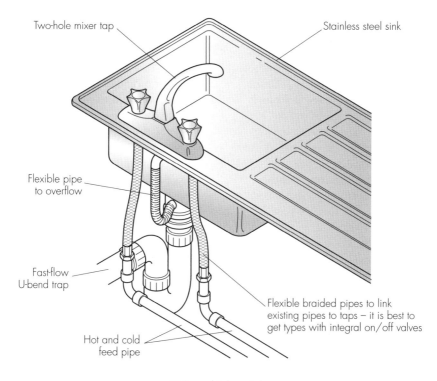

Two-hole mixer tap

Stainless steel sink

Flexible pipe
to overflow

Fast-flow
U-bend trap

Flexible braided pipes to link
existing pipes to taps – it is best to
get types with integral on/off valves

Hot and cold
feed pipe

Fitting a kitchen sink.

Fitting a kitchen sink

Let's say that you want to replace an
existing kitchen sink. You intend to
remove the old sink and fit a new one
that you have purchased as a flat-pack
kit that includes a cabinet, stainless-
steel bowl with a drainer at each side,
two-hole mixer tap, and all the parts
that make up the U-bend trap.

Method

1 Having made sure that you are
   stocked up with enough prepared
   meals for the day, turn off the
   mains water and use a hacksaw
   to cut the old sink away from its

supply and waste pipes – the
copper pipes for hot and cold
water and the plastic waste pipe
that runs through the wall into
the yard gully. If the waste pipe is
more than five years old, remove it.

2 Assemble the cabinet according to
   the directions, and clamp the bowl
   and drainers in place.

3 While the back of the cabinet unit
   is readily accessible, position the
   mixer tap unit in the two holes,
   making sure that it is centred, and
   then fit the back nuts and 'top-hat'
   washers on the two threaded tails.
   Clench the nuts up tightly.

4 Fit a flexible integral-valve connector to each of the two tap tails – one for hot and the other for cold. These are a great option for two reasons. They make the whole process of joining the sink to the existing supply pipes very much easier, and also allow you to service the taps without having to cut off the water supply to the whole house. Use a spanner to clench the connectors tightly.

5 Fit the U-bend trap to the waste outlet, with the flexible pipe banjo overflow running upwards to the overflow hole.

6 When the connectors and trap are in place on the sink, move the whole cabinet (complete with the sink unit) over to its final position and decide whether or not you need to extend, replace, or otherwise reshape the existing pipes for hot and cold supplies, and the waste pipe.

7 Take the sink unit out, fit the waste pipe and extend the pipes for hot and cold supplies as necessary. Move the cabinet back in place against the wall, and link up the pipes for hot and cold supplies and the waste pipe. Finally, wipe dry toilet tissue around all the joints and turn on the water. Note any leaks, make them good, and the job is done.

Fitting a washing machine

Washing machines are generally plumbed into the hot- and cold-water feed pipes in your kitchen or utility room, and the water runs out through

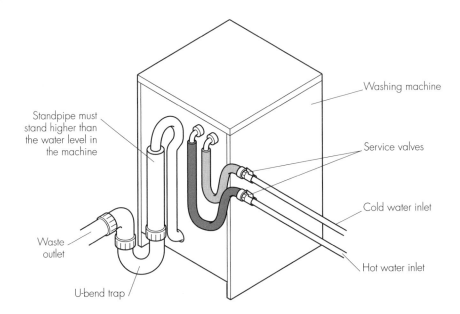

Washing machine fitting.

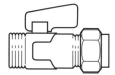

'Straight' inlet valve.

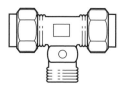

'T' inlet valve.

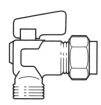

'Angled' inlet valve.

a 40 mm waste pipe (similar to the waste leading away from your sink). When you are considering where to put the washing machine, the biggest factor is where to run the waste pipe, because it needs to be fitted with a 6 mm fall in every metre run to ensure that the water runs away efficiently.

The easiest place to install a washing machine is in a cupboard space alongside your sink. In this location you will have easy access to the supply of hot and cold water as well as the sink waste pipe. Fitted kitchen cupboards are normally standard dimensions, so it's relatively easy to adapt a cupboard to take the washing machine – by cutting out the shelves, back and floor and making the kick board removable. You will usually find there is space to run the pipework behind the kitchen units.

Although most washing machines have two connections – one to be linked to the supply of hot water and the other to cold – it is possible to connect both fittings to the cold supply. The washing machine will then heat the water – an option that is slightly more expensive than taking hot water directly from the system.

Method

1   The washing machine needs to be connected directly into the hot- and cold-water feed pipes that run to the sink.
2   The easiest way is to use self-tapping (or self-boring) valves. Clamp the valve to the pipe and follow the manufacturer's directions for screwing the valve body into the clamp (a hole is drilled in the pipe and the unit is fitted in a single operation, without having to turn off the water). Connecting valves of this character can be bought in most DIY superstores and plumbers' merchants. The valves have a tap and a threaded connection for the washing machine hose. (See also 'Method for installing a dishwasher' on page 101.)

It is possible to connect one self-tapping valve to the cold-water feed pipe and then fit a Y-connector to link up to both the red and blue hoses from the washing machine, but it is better to fit two valves – one on the hot pipe and the other on the cold.

3   If you use ordinary valves, it is necessary to turn off the water, drain the system and fit branch pipes from the existing hot- and cold-water feed pipes.

4   There are several types of washing machine valve – angled, straight and T-shaped. Decide which is most suitable for your installation. A good option is to solder a 15 mm equal-T fitting and a length of copper tube into an existing pipe run, so that an extension leads towards your washing machine, and the washing machine hoses and valves are accessible behind the washing machine. Washing machine valves have a small red or blue tap that allows you to turn off the water to the washing machine, a thread to take the hose from the washing machine, and an integral compression joint that can be linked directly to a 15 mm hot or cold pipe.

5   There are several options for connecting the waste pipe. The easiest is to buy a proprietary washing machine kit that consists of a standpipe, trap and clips for connecting to the wall. The waste water hose from the washing machine must come up to a point that is at least two-thirds the height of the machine before it is linked into the waste standpipe. (The reason for this is that if the top of the pipe is not higher than the water level in the machine, the machine will simply siphon itself out and will never fill up with water.)

6   There are other ways of connecting to the waste pipe, depending on your particular situation. If, for example, there is already a waste pipe running behind the machine, you can fit a clamp-on connector directly to the existing waste pipe, so that the wastewater from the machine goes into the sink waste trap. The clamp-on connector is supplied in kit form complete with a hole-cutting tool that allows you to tap into the existing waste pipe. However, even if this type of fitting is used, not only will the waste hose from the machine still need to go to a high level, but the waste pipe will need to be linked to an anti-siphon fitting. (The standpipe option, on the other hand, does not need an anti-siphon fitting.) You can buy an in-line anti-siphon fitting that can be installed simply by cutting the waste hose at its highest point and pushing the two cut ends on to the anti-siphon fitting. Whichever set-up you choose, always read the manufacturer's instructions.

Dishwashers

Dishwashers are manufactured in two main sizes: large (which is the same size as a washing machine), and two-thirds size (which is sometimes easier to accommodate in a fitted kitchen). A dishwasher is very similar to a washing machine with regard to installation, but it only needs a supply of cold water because it heats the water itself.

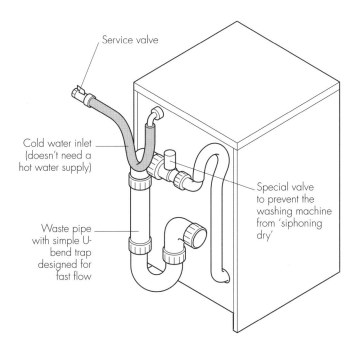

Service valve

Cold water inlet
(doesn't need a
hot water supply)

Waste pipe
with simple U-
bend trap
designed for
fast flow

Special valve
to prevent the
washing machine
from 'siphoning
dry'

Fitting a dishwasher (with sealed waste pipe).

The simplest place to install a dishwasher is usually in the cupboard to the left or right of the kitchen sink, where there is easy access to the water supply and drainage. It is of course possible to extend the drainage and supply pipes so that they run in the space behind the kitchen units, allowing the dishwasher to be located some distance away from the sink, but this may entail cutting the back out of several cupboards which can be a lengthy task.

The simplest way to tap into the water supply is to use a self-tapping valve, in just the same way as for a washing machine (see page 99). Read the manufacturer's instructions carefully before installing the valve.

Method

1 The dishwasher needs to be connected directly into the hot- and cold-water feed pipes that run to the kitchen sink.

2 The easiest way is to use self-tapping (or self-boring) valves. Clamp the valve to the pipe and follow the manufacturer's directions for screwing the valve body into the clamp (a hole is drilled in the pipe and the unit is fitted in a single operation, without having to turn off the water). You can, of course, install a normal washing machine valve instead (use one with a small blue knob) to connect into the plumbing under the sink.

3 If the dishwasher is to be some distance away from the sink, you may wish to solder an equal-T fitting in the cold-water supply to the kitchen tap, run a pipe behind the kitchen units to the dishwasher location and then end it with a small washing machine valve.

4 The dishwasher wastewater hose needs to be connected in a similar way to that of a washing machine. It must go to a level that is higher up than the water level in the machine (the hose must rise from the bottom of the machine to at least halfway up the height of the machine) before it hooks over or connects to the waste standpipe. If you want to connect the wastewater hose directly to the trap under the sink, you need to fit an anti-siphon valve into the wastewater hose at its highest point. Refer to 'Fitting a washing machine' on page 98.

5 Look at the waste trap under the sink – some have a small 'blank' that can be unscrewed, allowing you to fit a wastewater hose direct. It has an integral one-way valve that stops wastewater from the sink going into your dishwasher. If the trap does not have a connection point, you can easily replace it with one that does.

6 If the dishwasher is some distance away from the sink, you will need to install a 40 mm wastewater pipe behind the kitchen units, with a standpipe kit fitted on the wall behind the dishwasher. These kits are available from DIY stores.

Water softeners

If you live in an area with hard water, it will in time not only block up your pipes, cold-water storage tank, taps, showerheads and so on, but will also reduce the efficiency of your boiler and immersion heater. To find out whether you have hard water, look in your electric kettle. If you can see a chalky white coating of limescale, it is likely that your water supply contains calcium and magnesium, which is introduced when the water flows through bedrock before going to the water treatment works.

A water softener diverts the cold water from the rising main through a tank containing resin, which absorbs the calcium and magnesium ions and releases sodium ions, with the effect that your hard water problems are cured. The water softener machine periodically flushes the resin with salt to wash away the accumulated calcium and magnesium, and the wastewater runs into the drain in much the same way as the wastewater from a washing machine or dishwasher.

The average small domestic softener unit is usually designed to fit under a kitchen worktop, where it is connected to the rising main and the wastewater drain. The supply of drinking water to the kitchen cold tap does not go through the water softener, as the minerals in unsoftened water are beneficial.

If you are considering having a water softener installed but would like to have more information about what is involved, the Expert Points opposite will be helpful.

10 Expert Points

ALL YOU NEED TO KNOW ABOUT WATER SOFTENERS.

1 SALT
A water softener needs to be topped up with salt every few months or so.

2 PREVENTING CONTAMINATION
Water authority regulations insist that you fit a non-return valve between the water softener and the stopcock from the rising main, in order to prevent contamination of the drinking water supply.

3 DRINKING WATER
The cold-water tap at the kitchen sink is taken from the rising main at a point between the stopcock and the non-return valve – so that your drinking and cooking water remains untreated and you still receive beneficial minerals.

4 GARDEN TAP
If you have a connection to a garden tap, this should also be taken from the rising main between the stopcock and the non-return valve, otherwise you'll be spraying expensive softened water on the garden!

5 SERVICING
Water softeners sometimes need servicing, so it is better to install three valves as shown in the diagram on page 104. The valves allow you to turn off the water softener and open the bypass valve, providing a supply of unsoftened water whenever the softener is out of commission.

6 SAVINGS
Although a water softener is fairly expensive, this is offset by the fact that you will use less washing powder in your washing machine and dishwasher, your pipework and heating system will last longer, and your clothes and dishes will be subjected to less wear and tear.

7 DRAINCOCK
You will need to fit a draincock in the rising main just above the non-return valve, because the valve will make it impossible to drain water from any point above it.

8 STANDPIPE
On the wastewater side of the water softener, you need to install a standpipe (as for a washing machine) for the wastewater hose to go into.

9 WASTE CONNECTION
If you use a waste connection that links up directly with the trap under the sink, fit both an anti-siphon valve and a non-return valve to prevent dirty sink water being sucked back into the water softener

10 FUSED CONNECTION UNIT
The water softener must be connected to a fused connection unit (FCU) nearby, so make provision for this. This should really be done by an electrician. (An FCU isolates the appliance in the event that there is an electrical fault.) If you are purchasing the unit direct from the manufacturer, ask them if they can build the fitting costs into the price.

Anatomy of a water softener

A water softener needs to tap into the cold-water rising main as well as a drain, so the kitchen is usually the ideal place for installation. However, if the stopcock on the water supply to your house is in a different room, such as a utility room, it may be easier to install the water softer there.

As you can see in the illustration on page 104, the water softener comes in the form of a small tank, normally half the width of a washing machine, which fits under a kitchen unit.

A water softener does not make any noise in normal day-to-day operation, but when it flushes the resin compartment it will be pumping water down the drain, rather like a washing machine. When the water softener is flushing itself, it cannot treat water at the same time. This being so, it is important to make sure that the timer is set to flush the system in the early hours of the morning when softened water will not be required. Remember that if you have a power cut the timer will need resetting to the right time.

Plumbing in a water softener

If you have an indirect water supply system (see page 30), the water coming into your house through the rising main goes straight up to the cold-water storage tank in the loft, with a branch off to your kitchen tap (and maybe an outside tap in the garden). (See also page 30 for details of a direct water supply.)

A water softener diverts this cold-water feed through the softener treatment tank before returning it to the pipe that feeds the cold-water

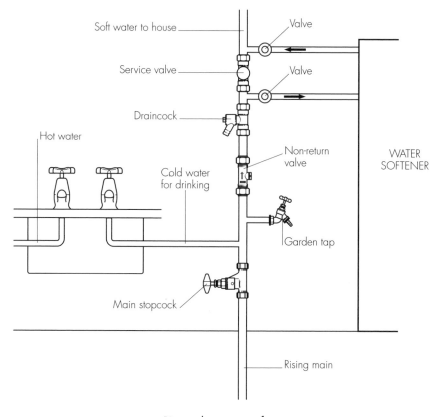

Pipework to water softener.

storage tank in the loft. This means that the water used in the majority of the house, including the central heating, hot water, cold water to bathrooms and the WC cistern, is all treated by the water softener. If you have the time and money, it is sometimes better to connect the WC cistern directly to the rising main before the softener, and fit a small nozzle in the ballcock valve in the WC cistern. (This is because toilets consume a lot of water during flushing, and although it may occasionally be necessary to use a descaler in the WC pan, the consumption of softened water will be considerably reduced by the implementation of this measure.)

Although it can seem a bit fiddly, plumbing for the water softener must be installed as illustrated, or there is a risk of water contamination.

Connecting up

Working from the mains stopcock (usually under the sink), fit a one-way valve with the flow direction heading into the house, a small draincock, a washing-machine-style valve for the feed to the softener, a service valve (fitted with a knob) to divert the water flow through the softener (and to bypass the softener if it is out of action), and another washing machine tap (the return from the water softener). Use valves that have a quarter-turn action, which are fitted with a knob that indicates whether the valve is open or closed. When servicing, this makes it easier to see at a glance whether the water softener is in flow or bypass mode.

Fitting a garden tap

A garden tap is very useful for jobs such as cleaning the car, watering the

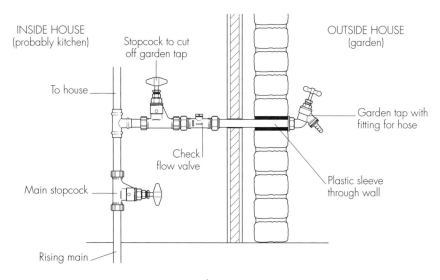

Fitting a garden water tap.

garden, washing the dog and filling up the paddling pool. Apart from 15 mm copper pipe and various joints to suit your situation, you need a stopcock, a non-return valve, a draincock and a bib-tap to complete the task.

Method

1 In one of the under-sink kitchen cupboards, use a drill bit 25 mm in diameter to drill a hole through the wall to the outside, at about knee height and as near as possible to the rising main.

2 Turn off the main stopcock and run the rising main dry. Cut into the main at a point directly below the drilled hole and fit a T-joint.

3 From the T-joint (and running directly up to the hole) fit a short length of pipe, then the stopcock, then the non-return valve, and then the draincock. Make sure that the arrows marked on the various fittings are all pointing in the direction of the flow.

4 And now for the difficult bit – making a neat job of the various joints while at the same time ensuring that the pipes fit closely to the walls. From the draincock, run a pipe through the wall so that it connects to the bib-tap, and screw the tap securely to the wall.

5 Finally, having first checked that all the joints are dry, stuff the hole in the wall with fibreglass roof insulation or foam, squeeze silicone sealant in and around the outside end of the hole to prevent the possibility of water leaking through, and the job is done.

Troubleshooting new appliance installations

If we discount all the electrical problems – bad wiring, faulty motors and such like – just about all the problems that relate to appliance installations have to do with the flow of water as it runs in or out of the machine. So for example the flow in to the washing machine or whatever might be too slow or too fast, or at the wrong pressure. And much the same goes for the wastewater running out. It might be so fast that there is a partial vacuum, or so slow that the machine senses that there is a problem and simply comes to a halt. And then again, there might be a leak or simply a kink in a pipe.

Appliance repairs

Is it under guarantee?

Most new appliances are under guarantee for 12 months; during this period the manufacturer is generally responsible if the device malfunctions. If you have misused the device in any way you may find that the warranty is invalid, so take care and only use appliances in accordance with the manufacturers' instructions.

Is it worth repairing?

Generally if you have purchased a top-quality item it will be worth repairing. The good news is that appliances like washing machines, tumble dryers, dishwashers and the like can nearly always be brought to good order. While an appliance like a modern washing machine will routinely last five years, with a bit of repair and maintenance its life could be extended almost indefinitely.

Cautionary notes

Most modern appliances are designed so that the manufacturer can tell if you have tried to take it to pieces. Manufacturers use unusual screws and fastenings as well as labels across casing joints so that they can tell whether you have been inside the appliance. If the appliance is still under warranty do not attempt to open the casing, as the manufacturer will certainly discover that you have done this and your warranty will be rendered null and void. While appliances like dishwashers can easily be repaired with a basic toolkit, other appliances such as washing machines will need specialist tools. It's a good idea to start out by having a good long look at the job to make sure that you have the tools for the task. Don't spend more money on specialist tools than the machine is worth.

The most common problems

Most of the problems with appliances have not so much to do with the appliance itself, but rather with the way it is fitted - meaning with the pipes and valves that govern and control the flow of water. If you have just fitted an appliance, then the best way forward is not to rush in and start stripping the machine down, but instead to look for the obvious.

● Appliance valves can easily be put on the wrong way round, impeding the flow of water. Always look at valves for flow direction arrows.
● It's easy when fitting an appliance to push it so hard against a wall that one of the flexible pipes is kinked, with the effect that the water cuts off.
● While self-boring valves are a great idea, there is just a chance that the small disc of metal that gets cut away will drift along the pipe and block the flow of water.
● Looped pipes can hold pockets of air that impede the flow of water - to the extent that the machine cuts out.

USEFUL TIPS FOR APPLIANCE REPAIRS

• In the context that most modern appliances are extremely complicated with hundreds of working parts, repairing appliances is best managed by a little-by-little procedure of working through a process of elimination – so that you can pinpoint the exact nature of the problem.

• Draw up a list of what has happened – how and when the machine failed, and all the things that the machine still does or fails to do. Try to narrow the problem down to a component shortlist – for example, a tap, valve or pipe.

• Many appliances are put together with exotic fittings – so that the washing machine or dishwasher is essentially 'tamperproof' and/or so that it can be easily assembled by an assembly line machine. It should be relatively easy to obtain a set of 'security bits' – meaning bits which will fit in a normal hexagonal screwdriver handle.

CHAPTER 6

Central-heating projects

Central heating

Although there are many ways of heating your house, the most convenient and effective is to install a 'wet' central-heating system. This system consists of a number of radiators around the house, usually at least one per room, and a boiler that uses either gas or oil to heat the water that is pumped to the radiators. Wet central-heating systems are popular because they can be easily turned on and off, unlike electric night storage heaters where the heat is stored overnight (Economy Seven) – and if the weather changes, the storage heaters will still be hot, even when you don't need any more heat!

There are two main types of wet central-heating system, although there are many variations in appearance and style. These systems are the traditional, open-vented system, and the sealed heating system.

CENTRAL-HEATING SYSTEMS COMPARED

SYSTEM	ADVANTAGES	DISADVANTAGES
Open-vented gravity-fed system. With the water being heated by a boiler the hot water moves by gravity circulation up to the hot water cylinder. An attic storage tank tops up the system.	The system is both easy to install and low-cost. It's a very common option to the extent that most plumbers are well able to make repairs and updates.	The system is such that the hot water cylinder ideally needs to be set as near as possible over the boiler. This need sometimes makes it a difficult option.
One-pipe system. With the water being heated by a boiler, the hot water is pushed around the system by a pump. A single large-bore pipe feeds the various radiators on the system.	The single large pipe that feeds the system makes for a very tidy layout, with pipe runs and untidy joints being kept to a minimum.	Sometimes the layout is such that the last radiators in line create a cold area that needs to be compensated for by extra-large radiators.
Sealed heating system. With the water being heated by a boiler, the system is topped up, via a non-return valve, directly from the mains.	The biggest advantage is the fact that the system is tank-free – so that it can be set up without the need for a tank in the attic or roof space.	The set-up is such that it involves high-quality components being fitted to a very high standard.
Hot air central-heating. Gas burner and electric fan circulate hot air. Plumbing not required.	Economical. Radiators are not required. Ceiling-height ducts and vents are discreet and easy to install.	Fan is noisy and the system is much less effective in rooms furthest away from the gas burner and fan.

10 Expert Points

THE FOLLOWING POINTS COVER THE
BASICS ABOUT CENTRAL-HEATING.

1 ASSESSING THE EXISTING HEATING SYSTEM
To establish which type of central-heating system you have, make a sketch and match it to the diagrams in this book.

2 LOOKING AT THE BOILER
Another way of checking which type of heating system you have is to make a note of the model number on the boiler and contact the manufacturer. Be specific – list the model, year of manufacture and any code numbers. Bear in mind that it is possible that the boiler has been part replaced; so, for example, while the casing might be original, just about everything else could be a replacement.

3 WATER CIRCULATION
In some houses, the water in the radiator pipes is circulated through the hot-water cylinder by gravity (convection), while in others a pump and electric valves are used to force water through the cylinder.

4 PIPEWORK
Most central-heating systems use 15 mm pipe to the radiators. If your pipes are much larger than this, you may have an outdated 'one-pipe' system, where the water is pumped around a loop and diverted to each radiator by convection.

5 SEALED CENTRAL-HEATING SYSTEMS
Sealed central-heating systems have an expansion cylinder that allows for the water to expand as it heats up. If you have this type of system, there will be a small pressure gauge somewhere on your central-heating boiler. When you are carrying out an inspection to establish whether this is the type of system you have, take care not to dislodge the electrical wiring.

6 MAINTENANCE
Central-heating systems are filled with water, which can cause rusting in the radiators. In addition, a black sludge gathers at the bottom of the radiators, reducing their efficiency. There are chemical additives available for cleaning out the system and for protecting it once it has been refilled. These additives are available from DIY superstores.

7 COMBINATION BOILER
If you have a gas supply to your house, it is usually easiest and cheapest to fit a combination boiler, which will both heat the radiators and supply all your hot water requirements.

8 CHANGING TO A COMBINATION BOILER
If you have an older system where the water is heated by one boiler and the radiators are heated by another, it is usually still possible to install a combination boiler that will do both jobs. The internet is a mine of information and possible sources for such a boiler. Just type in 'combination boiler'. Such boilers must be fitted by specialist gas fitters.

9 ADVANTAGE OF A COMBINATION BOILER
A combination boiler system does not require tanks and pipes in the loft. It is economical because you only heat the water you use. The downside is that, while the water coming from the hot tap is really hot, the flow is somewhat slow. This is a good set-up for a small flat.

10 REGULATIONS
Central-heating boilers can be very small, but there are still some rules regarding their positioning and the placing of the exhaust flue. Make contact with your local authority for advice on the regulations and their implementation.

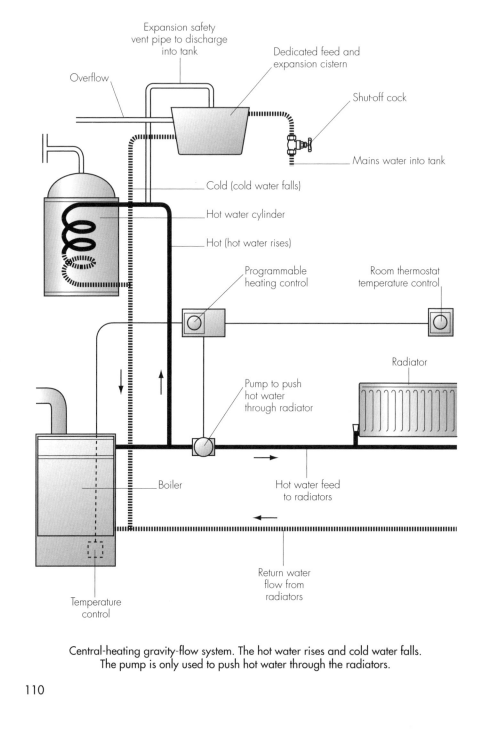

Central-heating gravity-flow system. The hot water rises and cold water falls.
The pump is only used to push hot water through the radiators.

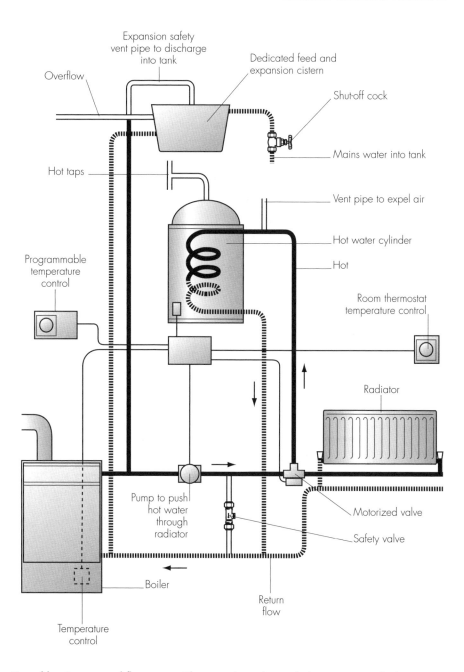

Central-heating pumped-flow system. The pump is used to push the water up to the hot water cylinder and through the radiators. The motorized valve controls the flow.

111

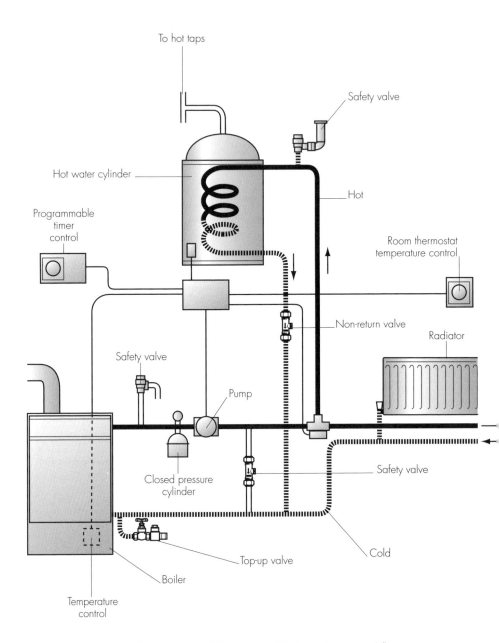

To hot taps

Safety valve

Hot water cylinder

Hot

Programmable timer control

Room thermostat temperature control

Non-return valve

Radiator

Safety valve

Pump

Closed pressure cylinder

Safety valve

Top-up valve

Cold

Boiler

Temperature control

Central-heating sealed-flow system. Similar to the pumped-flow system but does not need an expansion tank.

Open-vented central-heating system

Many homes have an open-vented central-heating system. In this type of system, water is heated by the boiler and pumped around the radiator circuit. The boiler can also be connected to the hot-water storage cylinder and may pump water through a heating coil in the cylinder. Sometimes the water is circulated through the cylinder by gravity without the need for a pump.

There is a small water tank in the loft, called a feed-and-expansion tank, which is easily recognized by its contents – rusty, muddy-looking water. This tank marks the highest point in the system. The tank keeps the system topped up, and allows for the expansion of the hot water in the system. The radiators and pipes only have a very low pressure in them.

There may be various devices in the system. If, for example, the design of the system means that it is prone to getting air pockets in the radiators, there will be an air separator to prevent this happening.

Sealed central-heating system

This particular type of central-heating system – using a modern combination boiler – provides hot water as and when you need it. There is no hot-water storage cylinder and no feed-and-expansion tank in the loft. Radiators can be installed anywhere in the house (it is not necessary for them to be on a lower level than a tank in the loft). It is a good option for a flat or house where there is no space available in the loft.

The boiler has an integral diverter valve. Water is heated in the boiler and circulated through the radiators via a pump. When a hot-water tap is turned on, the diverter valve diverts water through the heat exchanger.

Instead of having a tank to accommodate the expansion of the hot water in the system, sealed central-heating systems have a cylinder, called an expansion cylinder, with a rubber diaphragm inside. This allows for the water to expand as it heats up. There is a small pressure gauge on the boiler, and if the system becomes over-pressurized, some water is discharged outside via a safety valve.

Hot-water cylinders

Most houses have a hot-water storage cylinder, which is generally situated in the airing cupboard. In an open-vented central-heating system, there is a vented hot-water cylinder. Water is heated either directly (by a dedicated boiler or electric immersion heaters in the hot-water cylinder) or indirectly (by a coil called a heat exchanger in the hot-water cylinder, which is connected to the boiler for the central-heating system).

In a sealed central-heating system, there is an unvented hot-water cylinder. This is connected directly to the rising main and supplies mains-pressure hot water. Water is heated directly (by immersion heater) or indirectly (by a boiler).

If you don't have central heating, you will probably still have a hot-water cylinder where the water is heated by an immersion heater – a giant element resembling the one in an electric kettle. If you look on the hot-water cylinder, you will see a cable connected to a plastic cover – this is the immersion heater.

The indirect system is ideal for houses in areas where the water is not absolutely soft. If you live in a hard-water area and have installed a water softener, this reduces the risk of limescale build-up so it would be feasible to install a direct cylinder.

You can, of course, have an immersion heater electric element installed in the hot-water cylinder, even if the water is normally heated by a boiler. An immersion heater allows you to have hot water at the flick of a switch – a good option if you don't want to use the central-heating system or it is out of action.

Direct hot-water cylinder

The directly-heated hot-water cylinder is quite a simple system. Cold water is fed in at the bottom, where there is also usually a small draincock. The water is heated either by a dedicated boiler or by electric immersion heaters in the cylinder. A connection

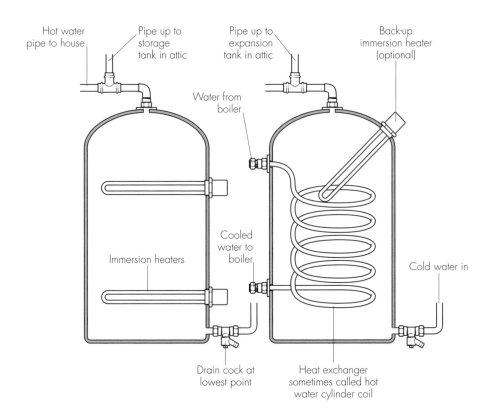

Direct hot water cylinder (left) and indirect hot water cylinder (right).

at the bottom of the tank feeds the boiler, where the water is heated. The hot water is then returned to the hot water cylinder through a connection at the top of the cylinder. The supply of hot water for the taps exits at the top of the cylinder through a T-joint; an expansion pipe called a vent pipe also runs from the T-joint to the cold-water storage tank in the loft.

Indirect hot-water cylinder

The indirectly heated hot-water cylinder has a heat exchanger (a coil) inside. Water heated by the boiler passes through the coil, and heat transfers to the stored water in the cylinder. The water in the coil is part of a self-contained system and is never mixed with the water in the cylinder that goes to the taps. This not only prevents limescale build-up in the boiler system, but it also extends the life of the system because it allows the use of anti-rusting additives in the boiler water.

The cold-water feed goes in at the bottom of the cylinder, where there is also usually a small draincock. The supply of hot water for the taps exits at the top of the cylinder through a T-joint; an expansion pipe called a vent pipe also runs from the T-joint to a feed-and-expansion tank in the loft.

Hot water from the central-heating system

In an open-vented central-heating system, the hot-water cylinder can be heated either directly or indirectly.

The direct method uses a dedicated boiler or electric immersion heaters in the cylinder. The indirect method works in conjunction with a boiler. As the boiler water circulates around the heat exchanger (a coil) inside the hot-water cylinder, the coil heats the water in the body of the cylinder, which is used to supply the taps. An indirect hot-water cylinder might also have an immersion heater for emergency back-up in case of boiler failure, but this is not employed as the day-to-day means of heating the water.

Direct boiler system

The direct system doesn't need a feed-and-expansion tank in the loft, but it does need an expansion pipe called a vent pipe that runs from the top of the hot-water cylinder to the cold-water storage tank in the loft. The boiler draws cold water directly from the bottom of the cylinder, heats it up and returns it to the top of the cylinder. There are thermostats in the boiler to ensure that the water temperature in the cylinder does not get too high.

There is a risk of limescale build-up, because the boiler and boiler pipework always have new water circulating through them. Chemical additives to prevent limescale cannot of course be used, since this water goes to feed the hot taps.

Direct boiler systems often have one or two electric immersion heaters in the hot-water cylinder for heating the water, and these elements are subject to limescale build-up. If there are two immersion heater elements, the lower one is often connected to an 'Economy Seven' electricity meter, and timed to switch on at night to take

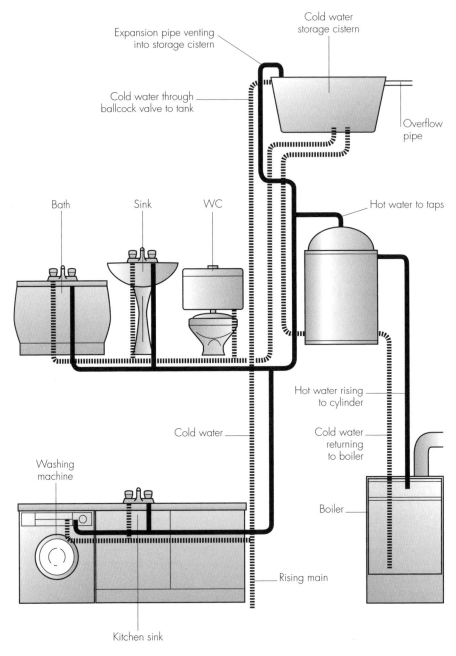

Expansion pipe venting into storage cistern

Cold water storage cistern

Cold water through ballcock valve to tank

Overflow pipe

Bath Sink WC

Hot water to taps

Hot water rising to cylinder

Cold water

Cold water returning to boiler

Washing machine

Boiler

Rising main

Kitchen sink

Direct hot water system.

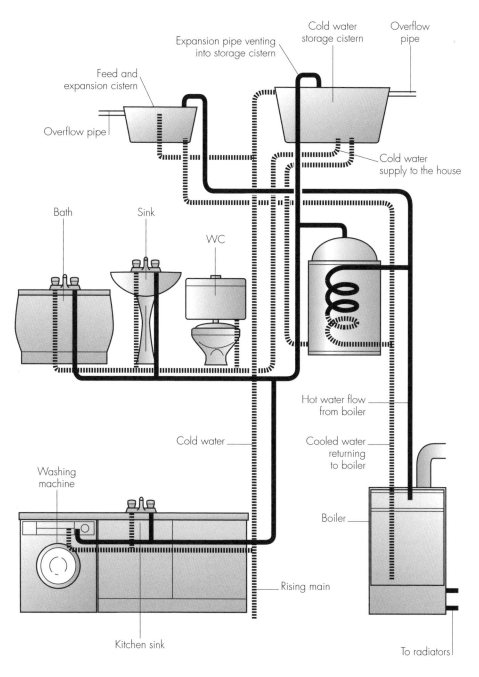

Cold water
storage cistern

Overflow
pipe

Expansion pipe venting
into storage cistern

Feed and
expansion cistern

Overflow pipe

Cold water
supply to the house

Bath

Sink

WC

Hot water flow
from boiler

Cold water

Cooled water
returning
to boiler

Washing
machine

Boiler

Rising main

Kitchen sink

To radiators

Indirect hot water and heating system.

advantage of cheap-rate electricity. The top element is normally connected to a pull-cord switch that is used to heat small quantities of water as required.

Indirect boiler system
The indirect system is the most common way of heating a hot-water cylinder. It features a central-heating boiler, a back-boiler on a fire, or a cooker of the Aga or Rayburn type.

The indirect boiler system has the advantage that the same water is circulating between boiler and hot-water cylinder (topped up by the feed-and-expansion tank in the loft), so that limescale build-up is negligible. The small feed-and-expansion tank (containing about 10 litres of water) in the loft keeps the heat exchanger in the hot-water cylinder fed and vented. (The cold-water storage tank in the loft cannot be used to supply and vent the heating system, since the water in the heating system is highly likely to have had anti-rust chemicals added to prevent the radiators and pipes from corroding.)

The water circulating around the boiler and the heat exchanger is called the 'primary circuit'. It is most important to remember that, while the vent pipe from the top of the primary system must go through the lid on top of the feed-and-expansion tank in the loft, it must finish just clear of the surface of the water and should never be submerged. The pipe must of course be open-ended. WARNING: It is important that you check the vent pipe from time to time to ensure that it is not obstructed.

Replacing a vented hot-water cylinder

If your existing vented hot-water cylinder has sprung a leak, is badly dented, or is uninsulated and letting out the heat, it needs to be replaced. Install a new, up-to-date cylinder of the same height, the same number of immersion heaters as the old one and a factory-fitted covering of polyurethane foam. New cylinders have a standard arrangement of holes to match the one you are replacing.

Method
1 Turn off the boiler (if you have one), switch off the electricity at the fuse box/consumer unit, and turn off the stopcock on the mains pipe. Turn on the bathroom taps and any draincocks, and run them until the system is dry.

2 Disconnect the immersion heater(s) from the electricity supply, and disconnect the various pipes so that the cylinder is standing free. Carefully ease all the feed and supply pipes out of the way and remove the old cylinder.

3 Take an immersion heater spanner (see note in Step 4) to the old cylinder and carefully remove the old immersion heater. Inspect the heater and decide whether or not to reuse it in the new cylinder or replace it with a new one. (As you have gone to all the time and trouble of fitting a new cylinder, it is a good idea to fit a new immersion heater.)

4 To fit the new immersion heater, first fit a new fibre washer, then

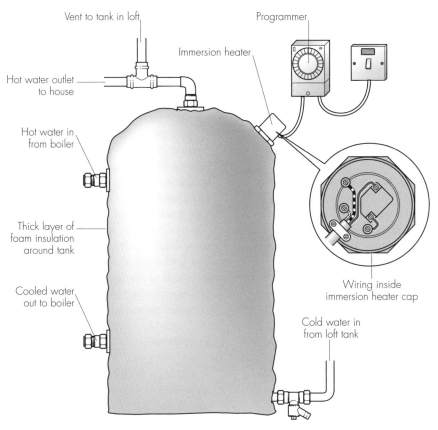

Vent to tank in loft

Programmer

Immersion heater

Hot water outlet to house

Hot water in from boiler

Thick layer of foam insulation around tank

Cooled water out to boiler

Wiring inside immersion heater cap

Cold water in from loft tank

Hot water cylinder.

A professional electrician needs to make final connections and test the wiring.

wrap 3–4 turns of PTFE tape tightly around the heater's thread, and finally screw the heater into the new cylinder. Note that while you can remove the heater from the old cylinder with a large, flat spanner, you will definitely need a large, box-type spanner to fit the new heater in the new cylinder. This being the case, you can either use the box spanner for both manoeuvres, or you can hire two spanners.

5 Carefully move the new cylinder into position, making sure that it is placed so that the various 'water in' and ' water out' holes are aligned with the appropriate pipes.

6 Check the condition of the draincock on the end of the cold-water feed pipe where it enters the bottom of the cylinder. If it has

119

seized up or if there simply isn't one, fit a new one.

7 Connect all the other pipes, wrapping PTFE tape around the threads. Reconnect the immersion heaters to the electricity supply.

8 Wipe lengths of dry toilet tissue around all the joints, close the draincock, open the stopcocks, turn off the bathroom taps and fill up the system. While the system is filling up, inspect the various joints for dampness, which will indicate a leak. If there is a leak, make good.

9 Finally, check out the whole system by turning on the immersion heaters and the boiler.

Unvented hot-water cylinders

In a sealed central-heating system, there is an unvented hot-water cylinder. This is connected directly to the rising main and supplies mains-pressure hot water to all the taps in the house. Water in the cylinder is heated directly (by immersion heater) or indirectly (by an in-built heat exchanger coil that is heated by a boiler). An indirect hot-water cylinder will usually also have an immersion heater to provide back-up in the event that the boiler fails or has to be shut down for servicing.

Unvented hot-water cylinders are less common than vented hot-water cylinders, even though the sealed central-heating system is an efficient system. Mains-pressure hot water is useful – especially for showers and for filling the bath quickly.

Expansion cylinder

As the water heats up in the cylinder it expands, and because there is no feed-and-expansion tank in the loft, an expansion cylinder must be fitted. This looks rather like a small gas cylinder and has a thick rubber diaphragm inside, which is pumped up with air. As the water expands it pushes its way into the diaphragm against the air pressure in the cylinder. Should the expansion cylinder fail, to the extent that there is no expansion space, the water will push past a safety valve into an overflow pipe. In addition, an 'over-temperature' valve must be fitted – this will also allow water into the overflow if the temperature of the water in the unvented hot-water cylinder becomes too high.

Regulations

There are many water authority regulations regarding the use of unvented cylinders. If you wish to install one, you will need to have the layout and final installation inspected by a qualified Institute of Plumbing engineer. With unvented hot-water cylinders, there is no need for a feed-and-expansion tank in the loft, since the water cylinder itself is pressurized directly by the rising main. It is the fact that the water is pressurized that most concerns the water authority, because there is a risk of contamination of the drinking water supply. This being the case, the most important component in this type of system is a one-way 'non-return' valve in the mains feed pipe.

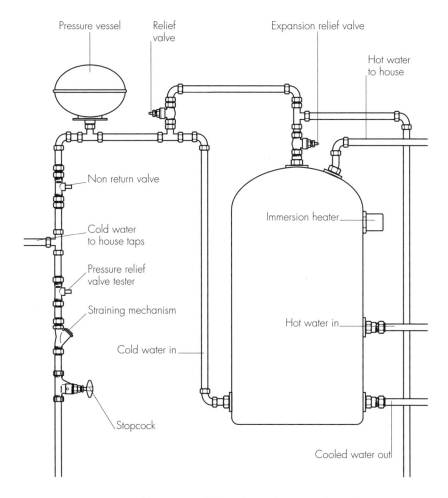

Pressure vessel   Relief
                   valve   Expansion relief valve

                                        Hot water
                                        to house

Non return valve

                                   Immersion heater

Cold water
to house taps

Pressure relief
valve tester

Straining mechanism

                                   Hot water in

Cold water in

Stopcock

                                   Cooled water out

An unvented hot water cylinder. The mechanical safety valves
and the overflow shown in the illustration must be fitted.

Because there is a chance that the overflow pipe from an unvented hot-water cylinder might discharge water, it must be run to the outside of the house. Some water authorities require two thermostats to be connected to the cylinder. The first one normally turns the heating system on and off; the second is an emergency shutdown device that shuts down the heating system if the water reaches 90°C.

Unvented hot-water cylinders are covered by building regulations, and require planning permission.

Thermal-store vented hot-water cylinders

The thermal-store cylinder is a good alternative to an unvented hot-water cylinder. It is easier to install and has fewer safety devices because the cylinder itself is not affected by mains water pressure.

Water is heated by the central-heating boiler and passes through the thermal-store cylinder, which contains a heat exchanger that transfers heat to mains-fed hot-water taps. A small expansion cylinder is fitted where the cold water enters the heat exchanger. This allows for the expansion of the cold water in the heating coil.

The thermal-store cylinder increases the efficiency of the central-heating boiler and provides hot water and central-heating at the same time. The advantage of the thermal-store

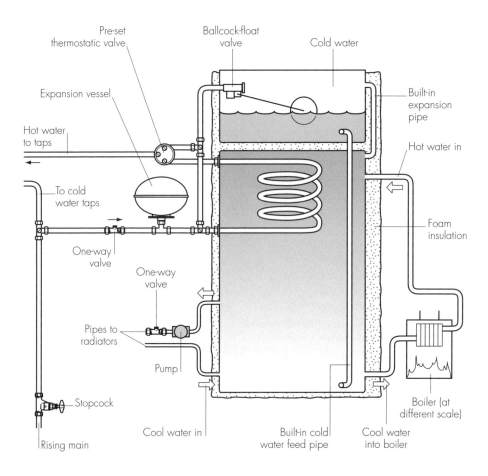

Thermal storage cylinder.

system is that the hot and cold taps both receive water at mains pressure – useful for running a shower and for filling the bath quickly! Because the water in the cylinder only circulates around the radiators and boiler, chemicals to prevent rust build-up can be added without ill-effect.

Feed-and-expansion tank

The thermal-store cylinder can be bought with or without a small feed-and-expansion tank built into the top. Bear in mind that this feed-and-expansion tank must be higher than the highest radiator, otherwise you won't be able to fill the heating system. If you need to have radiators higher than the thermal-store cylinder, you will need to fit a small feed-and-expansion tank in the loft, to ensure that there is sufficient head of pressure for the radiators.

Valves

The system is super-efficient, to the extent that the water supplied to the hot taps can be dangerously hot. So as a safety feature, it is normal practice to fit an automatic thermostatic mixer valve alongside the thermal-store cylinder: this adds a little cold water as the hot tap is flowing.

The system needs to have a one-way non-return valve in the radiator pipework, otherwise when the boiler is turned off, a thermal (convection) flow will be created that will empty the remaining hot water from the thermal-store cylinder – which in turn means the hot taps will not have any hot water feeding them.

Hard water

It is not a good idea to have a thermal-store cylinder in a very hard water area, as the heat exchanger coil and thermostatic mixer valve will accumulate limescale quite quickly. In areas with very hard water, you need to fit a water softener on the rising main as it enters the house. This will protect your entire system from limescale – see 'Water softeners' on page 102.

Central-heating boilers

Gas boiler with balanced flue

The installation of a gas boiler with a balanced flue is simple, but there are a few rules regarding the position of the vent on the outside wall (these tend to vary from one region to another, so obtain a set of regulations from the local council building control office).

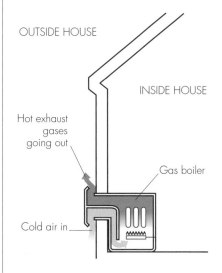

OUTSIDE HOUSE

INSIDE HOUSE

Hot exhaust gases going out

Gas boiler

Cold air in

Gas-fired boiler with a hole-in-the-wall balanced flue.

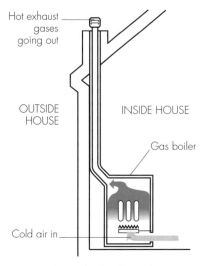

Hot exhaust gases going out

OUTSIDE HOUSE

INSIDE HOUSE

Gas boiler

Cold air in

A gas-fired boiler with traditional chimney flue.

A balanced flue through the wall has two passages. The outside air is drawn in, mostly using a fan, and then fed under the burner. The burner heats the boiler in much the same way as a gas cooker heats a kettle. The temperature of the boiler is controlled by various thermostats that make adjustments to the gas burner. The exhaust gases from the boiler pass through the flue to the outside, where they rise by convection up the side of the building. The fact that the hot gases rise means that you cannot fit a flue immediately underneath a window, gutter or balcony without any air space above it.

Oil-fired boiler

An oil tank situated outside the house supplies fuel to the boiler by gravity. (Or, put more simply, as the tank is higher than the boiler, the oil runs downhill to the boiler.) However, if the boiler has a fuel pump, the tank can be positioned at ground level or even buried. A device called a tiger loop is installed so that air can be removed from the pipeline.

In action, a motor-driven fan forces air under the boiler, fuel is injected through a very fine nozzle and an electric igniter lights the fuel jet, which in turn heats the boiler. Various thermostats control the temperature of the boiler by turning the supply of fuel on and off. The exhaust gases pass out of the top of the boiler and on into the chimney flue.

There are also oil-fired boilers with a balanced flue, which draw air in and pass exhaust gases out through the wall vent, in much the same way as a gas boiler of this type.

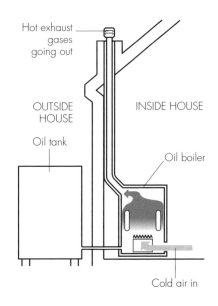

Hot exhaust gases going out

OUTSIDE HOUSE

INSIDE HOUSE

Oil tank

Oil boiler

Cold air in

An oil-fired boiler with traditional chimney flue

Fitting a new boiler

Gas boilers

A gas boiler is a good option –
relatively inexpensive to purchase,
cheap to run and generally trouble-
free. It must be fitted and serviced by
someone who has passed an approved
gas safety course run by CORGI
(Council for Registered Gas Installers).
Though this course only takes about
four days, it is not a worthwhile
option for most DIY-ers, especially
when you take into account the cost
of the tools and instruments required.

You can, however, fit all the
radiators, pumps and thermostats that
run off from the gas boiler, as long as
you follow the manufacturer's
recommendations. The important
thing to remember is that you must
not touch or tamper with the gas
supply pipe, valves, boiler and meter.

When you're looking for a boiler,
shop around and check out prices on
the Internet so that you know roughly
what each model costs. To check the
qualifications of a gas engineer and
make sure he or she is accredited by
CORGI, simply check the engineer's
registration number on the CORGI
website (www.corgi-gas-safety.com).

Oil-fired boilers

You can fit an oil-fired boiler yourself
as long as you follow the
manufacturer's instructions exactly
and check out the local building
regulations regarding the position of
the flue or the requirements for lining
the chimney. Chimney liners are easy
to install providing you have a head
for heights – you need to pull the liner
up the chimney from the rooftop!
The space between the liner and the
chimney brickwork is usually filled
with vermiculite granules to insulate
the flue. The liner and granules can
be purchased quite cheaply from
plumbers' merchants.

Fitting a gas boiler with a balanced flue

As a guideline – so that you know how
the system works – the illustration on
page 123 shows a typical installation
of a gas boiler with a balanced flue.
A 'combination' boiler is slightly
unusual, in that it heats the water 'as
required' without the need for a
header tank. The cold-water mains
feed is connected directly to the rising
main in such a way that the water is
heated and supplied the very moment
the hot taps are turned on. There is a
hot-water connection from the boiler
that goes to every hot tap in the
house. The feed pipes and return
pipes of the heating system are
connected to the radiator circuit, and
a pump in the boiler forces the water
around the circuit. The boiler has a
connection to the gas supply. There
are a variety of electrical connections
– a permanent 240-volt supply that
drives the fans and pumps within
the boiler, and a number of smaller
connections that provide room
thermostat control and heating timers.

Boiler maintenance

You cannot service a gas boiler
yourself unless you have passed a gas
safety course run by CORGI and have
the correct tools and test instruments.
Find a CORGI-registered gas engineer

125

to service the boiler safely. Make sure you get a quote before work commences. Once the boiler has been serviced, the engineer must give you a safety certificate for the boiler. Have the boiler serviced once a year.

You may like to arrange a service contract for your gas- or oil-fired boiler. There are various schemes available, such as service-only contracts, or a contract that will cover repair, labour and replacement parts.

Solid-fuel boilers (burning coal or coke) have a flue or chimney that needs to be swept twice a year by a chimney sweep, otherwise soot will build up in the flue and eventually stop the system working.

General maintenance to do yourself
To prolong the life of the central-heating system, the most effective maintenance you can carry out yourself is to make sure that the system is properly filled with corrosion inhibitor. This is important, as the components of the system (boiler, pipes and radiators) will not last long without protection. To find out whether or not your system has protection, do the following test. Draw off a water sample from one of the bleed taps at the top of a radiator when the system is on and the water is hot (take care not to burn yourself). Put the water in a jam-jar and drop in a shiny new steel nail. Fill another jar with ordinary tap water and place a similar nail in it. After a couple of days, the nail in tap water will go rusty as corrosion begins. On the other hand, the nail in the radiator water

should remain bright. If it does, your system doesn't need to have corrosion inhibitor added, but if it starts to go rusty, you need to add an inhibitor.

If the water bled from a radiator is black and murky, this is called sludge. The system needs to be flushed out with a proprietary chemical. To do this, tie the ballcock in the feed-and-expansion tank to a stick set across the top of the tank, so that water cannot flow into the tank. Drain the system by opening the draincock at the lowest point in the radiator pipework. Pour the flushing chemical into the feed-and-expansion tank and then allow the ballcock to partially fill the tank, at which point the flushing chemical will be drawn into the system. While some products require slightly different procedures, the next step usually involves draining the system and washing out all the flushing chemicals and sludge. Finally, refill the system with water and corrosion inhibitor to protect everything from rusting, and the job is done. It sounds a bit complicated, but in fact is quite easy to do. You won't go far wrong if you follow the instructions supplied by the manufacturer of the flushing chemical.

Radiators
At one time there were only two types of radiator – a huge, cast-iron radiator that looked like something out of a battleship, and a thin, dimpled-looking radiator made from pressed steel. Now there are radiators in just about every shape, size, design and material that you can imagine. There are single-panel radiators, double-

panel radiators, single-panel radiators with fins at the back, skirting radiators, radiators with integral electric fans, stainless-steel designer radiators that look like something out of a spaceship, continuous pipe heaters (really radiators that can be located under the floor or in a ceiling space), radiators that fit in trenches under the floor, and so the list goes on. Do not worry if you are somewhat bewildered, because the Expert Points on page 129 will help you decide what type to have.

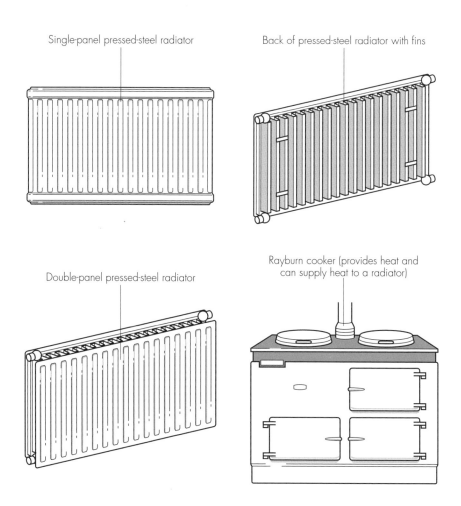

Single-panel pressed-steel radiator

Back of pressed-steel radiator with fins

Double-panel pressed-steel radiator

Rayburn cooker (provides heat and can supply heat to a radiator)

Types of radiator, heating options and heated towel rails (*see also* overleaf).

Low-level pressed-steel radiator
(for below a window)

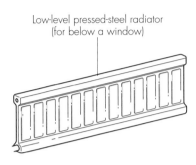

Combined pressed-steel
radiator and heated towel rail

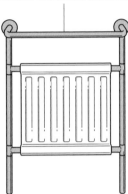

Modern 'designer' radiator
and towel rail

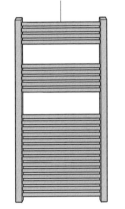

Old-style cast-iron radiator

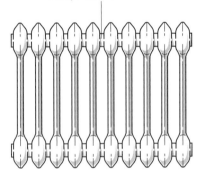

Skirting convector radiator

Heated bathroom towel rail

Fan-assisted radiator

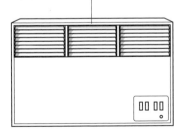

10 Expert Points

CONSIDER YOUR HEATING NEEDS BEFORE YOU CHOOSE WHICH TYPE OF RADIATOR TO INSTALL.

1 WHAT ARE THE CRITERIA?
Do you, for example, want to fit Edwardian-type cast-iron radiators? Or do you want to fit radiators at the lowest possible cost? Or do the radiators have to conform to some sort of design need – such as being flush with the floor?

2 POSITIONING
It was once considered best practice to position radiators below windows, because they were deemed to be cold spots. However, if your home is double-glazed, the radiators can be positioned to suit design and space needs since the areas surrounding the windows will be just about as warm as the other walls in the house.

3 WINDOWS
Be wary about placing radiators opposite a window – there is a possibility that they will draw cold air from the window and set up a cold airflow that cuts across the room. If possible, try to position the radiators at right angles to the window on one of the side walls.

4 SIZE AND NUMBER
The introduction of double-panel and finned radiators means that you can reduce the size and number of radiators in a room to the minimum. This is a good option if your room is long and narrow.

5 CALCULATIONS
To a great extent, the number and type of radiators relates to the size of the space being heated, so calculate the cubic capacity of the rooms that you want to heat. Find out the floor area by multiplying the width of the room by its length, and then multiply this by its height. For example, for a room 3 m wide, 4 m long and 2 m high, the sum is 3 m x 4 m = 12 sq m x 2 m = 24 cu m. You will need to heat 24 cu m. Use a Mears wheel to calculate the number of radiators required (see page 132).

6 UNDERFLOOR HEATING
Underfloor heating is expensive to install, but the water in the system doesn't have to be heated to the same high temperatures as with a radiator installation, so running costs are lower.

7 INSTALLING UNDERFLOOR HEATING
Underfloor heating can be installed in existing houses in several different locations – under concrete slabs, or suspended under wooden floors and/or in ceiling spaces.

8 TRENCH HEATERS
These are just small radiators that are set in trenches so that they are flush with the floor. They are a good option when you have floor-to-ceiling windows, and want to achieve a clean, minimal look.

9 CONVECTOR HEATERS
Wet central-heating convector heaters are an interesting option, and very good if you want instant heating. The heaters are made up of one or more finned pipes, rather like a larger version of a car radiator. In action, the cold air passes through the fins, heats up and rises, with the effect that hot air comes out of the top of the radiator and cold air goes in at the bottom. Some models are fitted with electric fans and dampers that allow you to turn the heat up or down to suit your needs.

10 SKIRTING RADIATORS
These are a good option when you want to achieve an even, all-round background heat in a room that is also heated by an open fire. Skirting radiators are readily available in kit form.

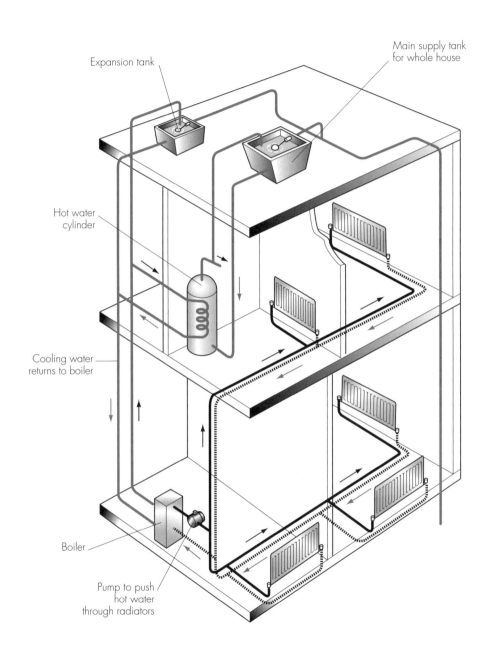

Expansion tank

Main supply tank
for whole house

Hot water
cylinder

Cooling water
returns to boiler

Boiler

Pump to push
hot water
through radiators

A basic heating and hot water system layout.

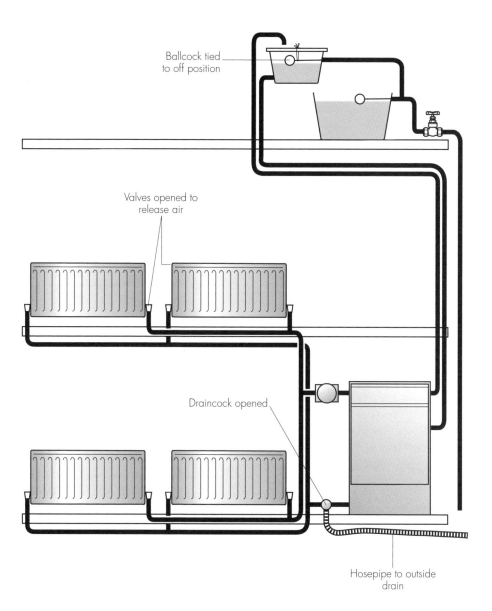

Ballcock tied
to off position

Valves opened to
release air

Draincock opened

Hosepipe to outside
drain

Draining the heating system.

Planning a new radiator system

Radiators are available in all manner of designs, lengths and forms. The standard type is made from thin, pressed steel and may have fins on the back to convect heat away more quickly. Double-panel radiators are used in larger rooms or situations where even more heat is needed. If you like the old-fashioned look, traditional cast-iron radiators are still being manufactured.

Mears wheel

To plan the system, carefully measure each room and note the size of the windows, the number of outside walls, the construction of the building and the function of the room. This information is used with a Mears wheel (which can be hired from a supplier of central-heating equipment) to calculate the size and number of radiators required.

The following guide suggests comfortable temperatures for different areas in the house.

COMFORTABLE TEMPERATURES

- Living room: 21°C (70°F)
- Bedroom: 16°C (60°F)
- Bathroom: 23°C (72°F)
- Kitchen: 16°C (60°F)
- Stairs: 18°C (65°F)

Radiator position

While it used to be common practice to install radiators underneath windows, to a great extent this practice has been rendered unnecessary by the fact that most houses are now fitted with double-glazing. If your house has double-glazed windows, the radiators can be installed on just about any wall that suits your needs and the design and furnishings of your home. (Avoid placing radiators directly opposite a window – refer to Expert Point 3 on page 129.)

Pipework

The pipework for radiators can be run under wooden floors, but usually the quickest method is to run pipework around the top floor (in a house that has two storeys or more) and drop pipes down the walls to the radiators on the lower floor. This layout means that you don't have to do anything to the lower floor. As this design involves creating 'inverted loops' in the pipework, so each radiator on the lower floor will need to have a draincock installed at its lowest point.

Radiators

If you don't have enough wall space to install long radiators, recalculate the requirements using a double-panel finned radiator that has an increased heat output for its size. Old-style cast-iron radiators don't have much surface area, so you will either need to install bigger ones, or raise their temperature by increasing the temperature of the water running around the system. If you want to fit cast-iron radiators, ask your supplier for the heat output information, so that you can plan your heating system accordingly.

Radiator valves

Hand-operated valves

The hand-operated valve is really just a stopcock that is set between the pipe run and the radiator. If you look at your radiator, you will see that there is a handwheel-operated valve at one end and a lockshield valve (covered with a plastic cap) at the other. The handwheel valve is used simply to turn the flow 'on' and 'off' to suit your needs – so if the room becomes too hot you would turn it off. The lockshield valve is set according to the radiator's position in the system, and then left well alone (it is a one-off setting made when a central-heating system is first installed).

Thermostatic valves

A thermostatic radiator valve can be fitted instead of the standard hand-operated valve. Thermostatic valves are relatively expensive to buy, but save money in the long term because they can be used to adjust the heat in the different areas of the home rather than having a constant heat throughout the house. For example, you might want low heat in the kitchen, high heat in the bathroom, negligible heat in the bedroom and moderate heat in the living room. You simply set the valve to the required temperature, and it will automatically cut in and out to keep each room at the desired temperature.

Radiator husbandry

Draining the radiators

1   Turn off the central heating. Once the water has cooled down, switch off the pump and shut down the two stopcocks to the two tanks in the loft (the cold-water supply tank and the feed-and-expansion tank).

2   Go to the boiler and locate the draincock at the lowest point in the system. Push one end of a garden hose securely onto the draincock and run the hose out into the garden or into a drain.

3   Open the draincock with a spanner. When water stops running out of the hose, go up to the top of the house and then, working your way down from the highest radiator in the system, open the bleed valves.

Cleaning and refilling the system

1   Once the system is empty, introduce a proprietary descaler into the system as recommended by the manufacturer. This will usually involve turning off the draincocks and bleed valves, pouring the descaler into the feed-and-expansion tank in the loft, running the boiler and then flushing the system.

2   Close the draincocks and radiator valves and open up the stopcock to the feed-and-expansion tank in the loft. When the system is full of clean water, go to the lowest radiator in the system and bleed the valve to let out trapped air. Gradually work up towards the topmost radiator in the system.

Cleaning (or replacing) a radiator

1   Roll back the carpet by the radiator and cover the surrounding floor

133

with old towels. Run a path of plastic sheet from the radiator to the door to the garden, so that you can carry the radiator outside without dribbling water across the floor covering.

2 You can use your hands to turn off the control valve at one end of the radiator, but you will need an adjustable spanner and an open-ended spanner to turn off the lockshield valve at the other end.

3 With a couple of bowls, a bucket and old cloths at the ready, and with a helper close at hand, go to the control-valve end of the radiator and use the two spanners to disconnect the radiator from the valve, all the while being ready to guide the water into a bowl.

4 When the water has stopped flowing out of the control-valve end of radiator, unscrew the connection at the lockshield-valve end, and lift the radiator from its bracket. Carefully take the radiator from the room and carry it out into the garden.

5 Wash out the old radiator with the garden hose, wind wraps of PTFE tape around the threaded tails, refit the connections, open the valves and bleed the air vents, and the job is done. Alternatively, replace the radiator with a new one of the same size.

Cleaning the pump on the boiler

1 Turn off the power at the main fuse box or consumer unit.

2 Go to the boiler, use a screwdriver to remove the cover plate from the pump, unscrew the connections and disconnect the wiring.

3 With a bowl, bucket and old cloths to hand, turn off the service valve taps at either side of the pump, undo the connections with adjustable spanners and lift the pump clear. (If there are no service valves, refer to 'Method for Replacing a Pump' on page 139.)

4 Take the pump into the garden and use the hose to flush out all the sludge. When you have cleaned it thoroughly (or purchased a new one of the same size and specifications), fit new fibre washers and use a spanner to fix the pump back in place.

5 Turn the valves back on, reconnect the electricity and run the boiler. Finally, use a screwdriver to bleed the large screw valve at the centre of the pump, and the job is done.

Balancing the radiators

The object of this exercise is to ensure that the flow of water through the radiators is balanced – by adjusting the individual valves – so that all the radiators reach the required heat. (Or, put another way, if you simply open all the valves to their full extent, the flow of hot water from the boiler will find the swiftest route round the system, with the effect that some radiators will run very hot while others stay cold.) You will need a pair of radiator thermometers, which can be hired from a tool-hire shop.

1 Turn off the heating system. When it has been off for a few hours and the water is cold, go to the

radiators and turn all the valves full on – the hand-operated valve at one end of the radiator and the spanner/pliers-operated lockshield valve at the other end.

2 If you now turn the boiler on and feel the radiators in turn, you should be able to work out the order in which the radiators heat up, and the direction the water flows in and out of the radiators.

3 Use tabs of masking tape to label the radiators with the order in which they heat up.

4 The next operation will take some time. Take the pair of thermometers and clip them to flow and return pipes just below the valves. Now, starting with radiator number one, close the lockshield valve and then open it little by little, until the temperature on the 'flow' (the hot-water-in end of the radiator) has reached a point about 10–11°C hotter than the temperature on the 'return' (the cold-water-out end of the radiator).

5 Continue making adjustments to all the radiators in turn until you get to the end of the circuit. If you have got it right, the valves on radiator number three will be slightly more open than those on radiator number two, and so on up the list, to the point where the valve on the last radiator in line will be almost full on.

Central-heating pumps and control valves

Nearly all central-heating systems have a pump that pushes the water around the system, sending it through the radiators and then back to the boiler. The design of central-heating pumps ensures that they are extremely quiet and unobtrusive. If you have quite a large house, especially a large tall house spread over a couple of floors, there will almost certainly be more than one pump – for example there may be one in the boiler and one in an upstairs airing cupboard.

Most homes have at least one control valve in the radiator system, fitted to ensure that the house is heated efficiently. The control valves are electrically operated by a central-heating timer, and you set them to go on and off as required. In indirect central-heating systems, a special valve called a three-port control valve is used to divert water from the central-heating circuit through the heat exchanger in order to provide the household with a constant supply of hot water.

If a valve or pump is not operating properly, some part of the system will not heat up. You will usually be able to locate the fault by a simple process of listening, feeling and elimination. Even the quietest valve and pump does make a slight noise and/or resonate when it is operating, so if you listen to the suspect item very closely and touch it lightly with your fingers, while at the same time getting someone to alter the central heating controls, you should be able to tell whether or not it is working. The Expert Points overleaf will help you track down the most common problems when they occur.

10 Expert Points

TROUBLESHOOT ANY GLITCHES IN
YOUR PUMPS AND VALVES BY
REFERRING TO THESE POINTS.

1 COLD RADIATOR
If the pump sounds like it's running, but one radiator or more stays cold, there may be an airlock in the pump. Find the pump and identify the bleed screw. Put a bowl under the pump and open the bleed screw to allow the air out. As soon as water appears at the screw, tighten it up and test the system. It is always possible that the small feed tank in the attic is running dry and letting air into the system.

2 ROTOR JAMMED
Sometimes the rotor in the pump will stick, and although you can feel a little vibration in the pump, the actual mechanism will not be turning. Some pumps have a cover screw that can be removed to access the spindle of the pump. Turn the spindle with a screwdriver until it is moving freely and test the system.

3 PUMP TYPES
There are two types of circulating pump: single-speed (fixed-head), and variable-speed. While a variable-speed unit is slightly more expensive, it is considerably more flexible than a single-speed pump and well worth the extra cost.

4 REPLACEMENT PUMP
If you need to buy a replacement pump, make sure it is exactly the same dimensions as the old model. If possible, take the old pump with you to compare it.

5 ADJUSTING PUMP SPEED
To adjust the speed of a central-heating pump, there is usually a small rotary switch. Some pumps require you to remove a cover or use a very small screwdriver or Allen key.

6 VALVE REPLACEMENT
Control valves are electrically driven. Before replacing a valve, check that the control unit, power supply and wiring to the valve are all okay.

7 SERVICE VALVES
Some central-heating pumps are fitted between two large service valves. This is very useful in that you don't have to drain the entire central-heating system to take the pump out. You will see a large screw set into the fitting either side of the pump. Turn the screws through 90° so that the slot is running across the width of the pipe – like a gate – and then remove and change the pump.If you are having a new boiler fitted it's well worth making sure that the pump is fitted between service valves, since this will make your life easier in the long run.

8 IMPELLER
Sometimes the pump impeller will jam on a bit of pipe scale and the system will fail to heat up. Before deciding to replace the pump, have a quick look to see if this is the case. If it is, remove the debris and refit the pump. Make sure that the entire system is completely free of debris before a new pump is fitted.

9 SAFETY
Before replacing a central-heating pump, it is essential to make sure that the power is fully disconnected. It must be turned off at the consumer unit or fuse box.

10 TOOLS
If the central-heating pump is fitted between two very large nuts, you will need two large adjustable wrenches in order to change it. Make sure that your wrenches open up to the required size. Whenever you are using long-handled wrenches take care not to overtighten the nuts, since this will cause problems in the future.

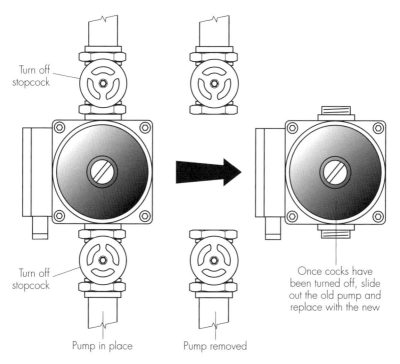

Turn off stopcock

Turn off stopcock

Once cocks have been turned off, slide out the old pump and replace with the new

Pump in place

Pump removed

Exchanging a pump.

Central-heating pumps

- Central-heating pumps rotate in one direction and have a flow arrow marked on the casing. Before you remove an old unit, make sure you know which way the flow is – which end the water comes in and which end it flows out – and use a marker pen to label it accordingly.
- Under the lead on the control box there is a screw connector where the electrical power is connected from the central-heating control. There may also be a dial or screw for adjusting the speed of the pump.
- On some units, you will find a metal or plastic cylinder about the size

of a salt cellar. This is a smoothing capacitor that prevents the pump motor from interfering with television and radio signals.

The motor should make almost no discernable noise, although if you put your hand on it, you should be able to feel a slight vibration.

- The large screw on top of the pump is usually the bleed screw for letting air out of the pump. Central-heating pumps are not self-priming, so if air leaks in, they won't pump until you bleed it out. You can do this by opening the screw slightly until it stops hissing and then tightening it.

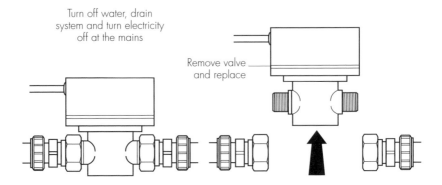

Turn off water, drain
system and turn electricity
off at the mains

Remove valve
and replace

Replacing a two-port valve.

Central-heating control valves
- There are two types of electrically operated valve. There is a two-port valve that shuts it on and off, and there is a three-port valve, which is called a diverter valve. This valve diverts the water flow through the hot-water cylinder when hot water is required.
- Most central-heating control valves – sometimes called zone-control valves – have a flow direction marked on the body. Make sure you note the direction of the flow arrow before you remove the valve.
- A valve is a simple, long-lasting device, capable of decades of service, but after many years of opening and closing, the electrical components can wear out and the water seals may leak. Valves are not expensive, but if you think you need to replace one it is well worth testing the old one properly to make sure that the fault is not with the wiring or the central-heating control timer.

- In an emergency, it is sometimes possible to disconnect the wiring from a valve, make the cable safe by disconnecting it at the controller, and then manually opening the valve with a screwdriver or small spanner. This procedure will result in you having no zone control, but it is a good option in certain situations – for example, the start of the Christmas holiday, when the shops are closed and your bedrooms are freezing.

Replacing or adjusting a heating pump
- Before you replace a central-heating pump, make sure that you clearly mark the flow direction on the pipework. There are several models of pump, but you must make sure that the new one has the same measurement as the old one between the two threaded ports that connect to the pipework, and that the threaded ends are the same size.

- Some pumps have a variable speed control. This looks like a little rotary switch and is usually to be found just inside the connection box.
- Pumps are available in several qualities, and as with most things in life, you get what you pay for.
- Don't forget that a failed pump might have a small piece of pipe scale stopping the impeller from moving, so check to see if this is the case before replacing the whole unit.
- Make sure your replacement unit is rated with the same capacity (flow rate) as the old one.

Method for replacing a pump

1 Turn off the power at the main fuse box or consumer unit. Make a sketch of the wiring layout and terminals. Disconnect the electrical connections to the central-heating pump.

2 If the pump has service valves at either side, turn them off. If not, you must drain the system at its lowest point. Connect a hosepipe to the draincock at the lowest point in the system and run this to the outside of your house.

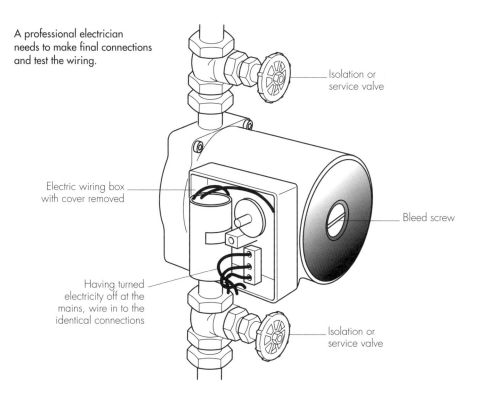

A professional electrician needs to make final connections and test the wiring.

Isolation or service valve

Electric wiring box with cover removed

Bleed screw

Having turned electricity off at the mains, wire in to the identical connections

Isolation or service valve

Rewiring a new pump.

3 Put a stick across the central-heating feed-and-expansion tank (if your system has one) and tie up the ballcock firmly, so that no more water feeds the system as you are trying to empty it.

4 Open the draincock and allow the system to drain. Put old towels around the pump and use two adjustable spanners to undo the large nuts and lift the pump clear.

5 Look inside the pump to see if the impeller is jammed with debris. If it is, clean it out, fit new fibre washers and put the pump back in again.

6 If you are replacing the pump with a new one, take the old pump when you go to buy the new one, so that you can get a perfect match. Fit the new pump and reconnect the wiring according to your sketch.

7 Close the draincock, put some corrosion inhibitor into the feed-and-expansion tank, and release the ballcock to refill the system. Bleed the air out of the pump and the radiators, and then restart the heating system.

Method for adjusting a heating pump and radiators

1 If the pump has a variable-speed switch, it can be used to increase the temperature of the radiators.

2 Alternatively, clip radiator thermometers on the two pipes at the ends of each radiator in turn and check the temperature. The difference between each thermometer should be 10–11°C. Refer to the section on 'Balancing the radiators' on page 134 for details of how to make any necessary adjustments.

3 If the radiators have a thermostatic valve on one end (slightly larger than normal radiator valves, with a scale printed around the cap), turn it to the hottest setting before balancing the temperature of the radiators. After each radiator has been balanced, you can adjust the speed of the pump so that the rooms reach the required temperature. If you have any doubts about your particular values, then ask the manufacturer of your radiators for an advice sheet

Heating controls

Both gas-fired and oil-fired central heating give you the ability (by means of a variety of control devices) to control the heating within your house to fairly precise limits. There are four main devices that can be incorporated into your system: room thermostats, timers or programmers, thermostatic radiator valves, and electrically operated zone-control valves.

Ideally, a house should be divided into zones and control devices fitted so that you only heat the parts of the house that you need. For example, it is not really necessary to heat your bedrooms during the day, or the sitting room in the middle of the night. It's a good idea to keep a diary or log of the various activities that go on in the house so that you know precisely how and when you use the space. You can then programme the different zones so that you make the most efficient use of your heating.

10 Expert Points

HERE ARE THE BASICS ABOUT THE VARIOUS HEATING CONTROL DEVICES AVAILABLE.

1 SUITABILITY OF CONTROLS

Automatic central-heating controls are really only suitable for use with gas-fired or oil-fired central-heating systems. Solid-fuel systems (coal or coke) are very slow to change temperature, so precise control is difficult. The best that you can do with a solid fuel system is to set the controls to deal with the predicted temperature.

2 TIMER PROGRAMMER

A timer programmer can be installed to allow you to adjust the time when the central heating and hot water come on and go off. It might have either a rotary-style clock timer mechanism, or a digital push-button display. Make sure that you position the programmer well out of the reach of any children in the household.

3 OPTIMIZERS

Special programmers called optimizers monitor the temperature in and around the house – inside and outside – and then adjust the system for the most efficient pattern of temperatures.

4 ROOM THERMOSTATS

A room thermostat or 'room stat' is usually mounted at about head height in the living room. It will switch the central-heating pump and boiler on and off to maintain the temperature in the room. Position it away from the coldest area of the room.

5 ZONE-CONTROL VALVES

If your house can be clearly divided into zones that can be shut down at various times during the 24-hour period, you can install a zone-control valve in the central-heating circuit. The valve can then be set to suit the way in which your household uses the house throughout the day.

6 DIFFERENT TEMPERATURES ACCORDING TO ZONE

You can add zone-control valves that are in themselves controlled by room thermostats. These allow you to have a different overall temperature in each zone.

7 THERMOSTATIC RADIATOR VALVES

Thermostatic radiator valves are a good, easy-to-fit option that allows you to adjust the temperature of individual rooms. The easiest addition to make to your system is to put thermostatic radiator valves in your kitchen and bedrooms, so that you can control the temperature in those rooms while the main room thermostat controls the boiler.

8 SERVICING THERMOSTATIC VALVES

When you are servicing thermostatic valves – changing washers, for example – always be careful that you do not cross the threads or over-tighten the cap-nuts. To this end, it is a good idea to use short-handled spanners so that leverage is kept to a minimum. If you have to use long-handled spanners, compensate for this by holding them about half-way along their length

9 POWER CUTS

If there is a power cut, make sure that the programmer is reset, otherwise the heating will come on and go off at the wrong time of day or night.

10 MINIMUM REQUIREMENTS

A single room thermostat will not really be sufficient to control everything. For example, if someone leaves the kitchen door or an upstairs window open, the room thermostat in the living room will not know about it. It is best to fit thermostatic radiator valves as a minimum. Modern thermostatic valves are now so widely used that they are beginning to come down in price and are no longer the expensive option that they once were.

141

Thermostatic radiator valve

Thermostatic radiator valves are a quick and easy way of controlling temperature room by room. They are usually designed to be fitted as a replacement for ordinary radiator valves, so you should find that the two compression joints line up nicely with the existing pipework.

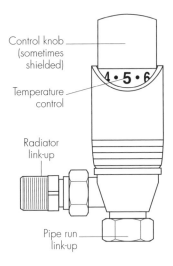

Control knob (sometimes shielded)

Temperature control

Radiator link-up

Pipe run link-up

Thermostatic radiator valve.

Thermostatic radiator valves have a flow direction marked on the body. When installing the valve make sure it is installed at the correct end of the radiator – the 'water-in' end.

The thermostatic valve needs to be fully open for refilling the system. There is usually a small plastic cap under the top of the valve that allows you to do this. If in doubt, look for guidance from the manufacturer's instruction and information sheet.

Timer programmer

A timer programmer is a digital or mechanical clock connected to a series of switches. The switches can be used to control motorized valves, the central-heating pump and the boiler. Using a combination of these controls, a timer programmer is able to give you both heating and hot water at precise times of the day and night. Some types of digital timer programmer can even give you different arrangements on different days of the week.

Programmers are quite tricky to wire up, so make sure that you study the manufacturer's instructions and have a clear understanding of just what needs to be done. As timer programmers are usually wired into mains-voltage electricity, it's a good idea to get your work checked by a qualified electrician or gas engineer before turning on the power. If you have any doubts, make contact with the manufacturer of the programmer, who will be only too happy to advise.

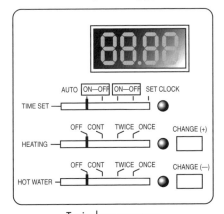

Typical programmer.
(A professional electician needs to make final connections and test the wiring.)

Replacing heating controls

If programming units or room thermostats fail, the likelihood is that they will need to be replaced rather than repaired. It is fairly easy to replace a programmer or room thermostat, as long as you carefully follow the manufacturer's instructions and diagrams. (If you have doubts about your work, get the wiring checked by an electrician before you reconnect the mains supply.) There is often a wiring diagram printed on the back of the old controller. It is always a good idea to make careful notes and sketches of the existing wiring before you dismantle anything – then you can at least put things back as they were.

WARNING: Always start electrical work by turning off the power at the main fuse box/consumer unit.

Method for replacing a zone-control valve

1 To change a zone-control valve, you'll need a good set of plumbing tools. Start by putting a stick across the top of the central-heating feed-and-expansion tank in the loft and tie up the ballcock so that water cannot fill the system. Once you have cut off the water supply, it is safe to drain all the water out of the heating system.

2 Make sure that you mark the flow direction of the valve on the pipe adjacent to the zone-control valve with a marker pen.

3 Put a hosepipe on to the draincock and run it out into the garden. Open the draincock and allow the system to empty.

4 Turn off the mains electricity supply to your house at the consumer unit and disconnect the zone-control valve's cable from the programming unit.

5 Use two large adjustable spanners to undo the compression fittings holding the motorized zone-control valve in place. If you cannot undo the fittings you will need to use a hacksaw to cut through the pipe a little way either side of the unit. (When purchasing the zone-control valve, try to get one of an identical size and type – it will greatly simplify the refitting procedure. If you cannot get an identical valve, it is best to fit about 200 mm of pipe to each end of the new zone-control valve and install this in the position of the old unit.)

6 Cut the pipework so that the new valve plus the additional length of pipe is a snug fit. Use either two compression joints or two solder joints to join your new valve to the old pipework.

7 Reconnect the wiring to the controller and turn on the power.

8 Close the draincock and untie the ballcock to allow the system to refill. At this point you can, if you so wish, add corrosion inhibitor to the feed-and-expansion tank. WARNING: Make sure that you pour the inhibitor into the small feed tank, NOT THE LARGE SUPPLY TANK.

9 Finally, bleed air out of the central-heating pump and the radiators, restart the system, and check that everything is working properly.

143

CENTRAL-HEATING INSTALLATION TROUBLESHOOTING

SYMPTOM	POSSIBLE FAULT	ACTION TO TAKE
Your oil-fired central-heating boiler keeps getting air in the feed pipework to the jet, making the flame start and stop or not light altogether.	In some oil-fired boiler installations the pipework design means that air gathers in the system that is feeding oil from the tank to the boiler.	Install a device called a 'Tiger loop'. This is a clever gadget that removes air from the oil feed pipe automatically, ensuring a smooth flow of oil to the boiler. It fits into the pipework just before it enters through your house wall.
Your radiators keep getting air in them and you find that you are having to bleed them every few weeks to keep them working.	This is a fairly common problem, usually caused by a loose pipework connection on the suction side of your central heating pump. It can also be caused when the header tank isn't filling up with water correctly.	Carefully check the pipework on the suction side of the pump leading towards the boiler, and carefully seal each pipework joint. You may need to fit a new ballcock in the header tank (if you have one) to keep a good level in the tank.
The bottom of a radiator is cold or warm while the top is hot, even after the radiator has been on for some time.	After a while black sludge caused by rusting within the radiator settles at the bottom. This restricts the flow of water in the bottom of the radiator while the top still remains hot.	Turn off the radiator valves, remove the radiator, flush the radiator out with fresh water and refit it. A corrosion inhibitor in the circulation water will slow down the rusting process.
You have fitted a new thermostatic radiator valve and it does not appear to work, even though the rest of the system seems fine.	The thermostatic radiator valve has been fitted the wrong way round, or on the wrong end of the radiator.	If you follow the pipework back to the pump surge you can work out which is the feed side of the radiator. The thermostatic valve has an arrow indicating flow direction. Refit the valve correctly.
The central heating pump is humming, as if working, but the water is not circulating through the radiators.	Sometimes scale from the inside of the pipes or rust from the radiators can jam the pump impeller. The motor is still trying to turn, which makes the humming noise.	Drain the system and remove the pump, clean out the pump and turn the impeller by hand to make sure it is working freely. If it still won't turn freely maybe a bearing has gone. Replace the pump.

SYMPTOM	POSSIBLE FAULT	ACTION TO TAKE
There is one radiator in your house that is always cold while the others are hot, and it seems there is no air in the top of the radiator.	Sometimes an airlock can occur in the pipework, especially when the system is refilled after a repair.	Turn off the central-heating system. Close all the other lock shield valves in the house, writing down the number of turns required to shut the valves so they can be reset later. Turn the heating on and let the pump force the water around the radiator. Reset all the lock shield valves.
You cannot achieve an even temperature in all the radiators in the house even after days of playing around with the valves.	It is very difficult to balance radiators correctly. Each lock shield valve must be correctly adjusted for the system to work properly.	Buy two radiator thermometers. Hang them on the pipes either side of the radiator. Fully open the thermostatic valve. Adjust the lock shield valve until temperature difference between the thermometers is 10–11 °C. Repeat for each radiator.
The pilot light in your gas boiler goes out as soon as you take your finger off of the relighting button, no matter how long you hold it in.	There is a safety sensor in gas boilers which will not allow the pilot light to stay on unless it is lit. The sensor can fail and cut off the pilot light gas supply.	This sensor is very cheap and will only take a gas fitter a few minutes to replace. The whole job from start to finish should only take an hour. You must use an approved gas fitter for this repair.
The hot water from your combination boiler is now only warm but the flow rate from the taps has increased.	There is a flow regulating valve in combination boilers which slows down the mains pressure water so that it only goes through the boiler at a speed that can be easily heated to the correct temperature. This valve needs to be replaced.	Get a gas fitter to replace the flow regulating valve. If the boiler is under warranty this repair should be free.
The pressure gauge on your combination boiler has been dropping slowly over a few weeks. The radiators are becoming less effective.	You have a very small water leak somewhere in your radiator system. This is probably so slow that it is evaporating before you notice a wet area.	Look carefully at all the joints in the system, especially the radiator valves. You should see white powdery deposits where the water is evaporating from the warm pipe. Cure the leak.

CHAPTER 7

Emergency repairs

Dealing with emergencies

In the context of plumbing, an emergency is a situation where there is a danger of accident to persons and/ or property. For example, a complete plumbing shut-down on a cold winter's night is unpleasant and uncomfortable, but cannot be compared with a situation where the tank in the loft gushes water over the main fusebox. (Of course, the definition of 'emergency' depends on individual circumstances, whether elderly people or young children are affected by the plumbing problem, and the consequent urgency of making a repair.)

If you are a beginner, a serious emergency scenario (such as water gushing out of the loft) is not really the time to put your raw DIY plumbing skills to the test. You should of course take action to prevent a problem worsening – such as turning off the water at the main stopcock – but if the situation is getting out of control, you will need to bring in a professional to deal with it.

Method and procedures

Water is both messy and destructive. One ill-considered mistake can be time-consuming and costly. This doesn't mean that you have to tremble and shake at the very notion of touching a stopcock; it simply means that you always have to follow the same methodical series of procedures

before you start work. Let the Expert Points opposite be your guide, but don't wait until a plumbing emergency happens before reading this section! You may need to act quickly in an emergency.

Dealing with electric shock

An electric shock can kill! If the victim stops breathing, he or she can die. You need to act fast, because precious minutes can make all the difference between life or death, but perhaps more than anything else you need to act coolly and methodically. If there are two people present at the scene, one person should immediately go to phone for an ambulance while the other stays with the victim.

The Expert Points on page 148 deal with a worst-case scenario, where only you and the victim are present. The victim has received a severe electric shock and fallen from the top of a step-ladder. He is flat out on the floor and still gripping an electrical appliance. His breathing seems to have stopped. It's important to follow points 1–10 in sequence. Of course, if you have worries about the very thought of such a worst-case scenario, then you have two options. You can always make sure that you are working as part of a team – with at least three people present – whenever you embark on any plumbing job that will involve coming into contact with electricity, for whatever reason, or you can read the relevant sections of this book and then cut your costs by managing a fully trained plumbing professional to do the work that needs to be done.

10 Expert Points

USE THESE BASIC GUIDELINES TO PREPARE YOURSELF FOR ANY PLUMBING SITUATION.

1 SAFETY
Never work when under the influence of drink or medication. If you have any doubts about the state of your health, ask your doctor for advice.

2 FITNESS
Make sure that you are physically up to the job. If you would like to try a challenging DIY task, but have doubts about your physical fitness, once again, seek advice from your doctor.

3 ASSISTANCE
It's a good idea to ask a friend or partner for help – to hold the torch, pass your tools and so on. If you have to work alone in the cellar or loft, at the very least tell a neighbour what you are doing. If possible, carry a mobile phone.

4 ADVANCE PLANNING
If you want to make changes to your plumbing over a long period, say over a series of weekends, you must plan accordingly and work out how to provide a supply of water for drinking and cooking, arrange to use a neighbour's toilet, and so on. At the very least, you will need a supply of fresh drinking water.

5 AVOIDING MISTAKES
Mistakes can be made if you are trying to work in a rush and/or you are over-anxious. Even when dealing with an emergency, consider all the options and take your time repairing it.

6 CHECKING THAT THE WATER IS OFF
Never take things for granted. You may have turned off a particular stopcock, but you must still make several tests (such

as turning on various taps and draincocks around the house) before you start sawing through any pipes.

7 TURNING OFF ELECTRICITY
Be very careful in situations where the electricity needs to be turned off and there are other people working in the house. Accidents will happen when one person is working and another person throws a switch or replaces a fuse without letting them know that they are about to do this. You should also make sure that everybody who is in the house while you are working knows what you are doing. If you have doubts, remove the appropriate fuse and put it in your pocket before you start working.

8 WATER-GUSHING EMERGENCY
In a water-gushing emergency, always start work by turning off the heating system and the main stopcock, and by turning the taps in the bath full on. In such an emergency the ideal is to finish up with all the stopcocks turned off, the system completely empty and a good supply of water left for drinking and flushing toilets.

9 PROBLEMS BEYOND YOUR CONTROL
If the water cuts off, do not automatically assume that there is a problem in the house. Ask your neighbours if they are experiencing a similar lack of supply, because it might be a problem with the supply pipe in the road. You don't want to waste time running checks on your own system if the failure is part of a larger problem that is beyond your control.

10 PROFESSIONAL CHECK
If you have replaced a pump yourself but have any doubts about the standard of your work, ask a trusted plumber to check it out before you start up the pump.

10 Expert Points

HOW TO DEAL WITH ELECTROCUTION.

1 DO NOT TOUCH THE VICTIM!
Go to the main fuse box/consumer unit and turn off all the power. If this isn't possible, use a wooden or plastic implement (such as a broom or walking stick) to knock the victim's hand free from the electrical appliance. Only as a very last resort, if everything else fails, attempt to pull him free by gripping a piece of loose clothing, such as a jacket. At no point touch his body.

2 SAFETY OF ONLOOKERS
Get any children and pets out of the room as quickly as possible.

3 POSITIONING FOR RESUSCITATION
Once the victim is clear of the source of the electrical current and the power is off, gently ease him over so that his head is slightly tilted back, and his tongue is rolling clear of his airway.

4 RESUSCITATION
If the victim still isn't breathing, pinch his nostrils together, place your mouth over his so that it covers it completely, gently blow until his chest rises and then remove your mouth. Repeat this procedure at intervals of about 5–6 seconds.

5 CALL AN AMBULANCE
After 10–12 repetitions, find a phone, dial 999 and call for an ambulance. Better still, send someone else to make the call.

6 ATTRACTING HELP
If there is no phone in the house, go into the street and shout for help.

7 CONTINUE WITH THE RESUSCITATION
Continue with the resuscitation procedure until the victim starts to breathe or until an ambulance arrives with paramedics.

8 PUT THE VICTIM INTO THE RECOVERY POSITION
Once breathing has started, ease the victim over on to his side, with his head resting sideways on his arm, and make sure that his tongue is clear of his airway.

9 KEEP THE VICTIM WARM
Cover the victim with blankets. DO NOT try to sit him up or give him a tot of rum or a cigarette, or anything else.

10 KEEP WATCH
If the victim vomits, gently wipe away the vomit and make sure that the victim's tongue is still clear of his airway.

Emergency repairs

As just about every householder will know, it's getting more difficult to find good, reliable, professionals. The Yellow Pages directory is full of plumbers who promise 'a quality service', declaring that they are 'the best that you can get', 'the cheapest', 'the most reliable', 'the fastest' and so on. However, when it comes to actually finding a plumber in an emergency, you will find that things get more difficult.

We did a little experiment to test out such claims by plumbers. On a rainy Saturday evening in winter, we phoned half a dozen 24-hour plumbers picked at random from the Yellow Pages, and asked how much it would cost for them to come over to stop a leak. Of the six calls, four said that it wasn't possible, one said something really interesting about sink plungers, and the remaining one said that he could do it for £100 for the

10 Expert Points

DEALING WITH AN EMERGENCY, SUCH AS WATER GUSHING THROUGH THE CEILING, REQUIRES YOU TO ADOPT A LOGICAL, STEP-BY-STEP APPROACH.

1 ASSESSMENT OF OTHER RISK FACTORS
The first thing to do is to assess possible risk factors aside from the gushing water – the baby, saucepans on the gas cooker, electric appliances, pets loose, aged relatives wandering about, accidents, and so on. Take steps to deal with all of these.

2 TURN OFF ELECTRICAL APPLIANCES
Make sure that all electric appliances are turned off, to prevent electric shock. Move computers and TVs out of harm's way.

3 DRAIN THE SYSTEM
To drain the system, you need to turn the central-heating boiler off, and turn the bathroom taps full on. Turn the stopcock under the sink off.

4 RETAIN WATER FOR DRINKING ETC.
While all the water is draining off, ensure that you still have a supply of water by filling up as many saucepans and buckets as you can find. Put a plug in the bath, but ensure that it doesn't overflow.

5 CHANGE YOUR CLOTHES
Change into comfortable working clothes and non-slip shoes.

6 ALTERNATIVE: LEAVE UNTIL MORNING
If you are tired and stressed, do not attempt to do anything now. Simply go to bed and return to the problem in the morning.

7 FIND THE SOURCE OF THE PROBLEM
Take a torch and have an exploratory look around the loft or roof space.

8 QUICK REPAIRS
Having found the problem, assess whether you can make a swift repair.

9 BIG PROBLEMS
If you see that part of the system (such as a pipe or tank) has fractured, wait until daylight and then either make the necessary repairs yourself or call in a qualified plumber to do the job for you.

10 MYSTERIES
If you cannot identify the problem, wait until morning and call in a professional. Never attempt to carry out any repairs unless you are absolutely sure what the problem is and that you can fix it.

first hour and £80 after that, with the cost of parts added on top. That may seem very expensive, but you have to look at the situation from the plumber's viewpoint: he is going out on a winter's night to sort out goodness knows what, his Saturday evening is ruined, and there is wear and tear on the van to take into account – perhaps £100 is justified.

So every householder should really work at a bit of basic plumbing DIY, so that he or she can at least ensure that the house has water and heat. At the very least, you ought to know enough about plumbing to get you out of difficulties. When a plumbing problem eventually occurs – and this might happen at any time of the day or night – you must be able to swiftly assess the situation, decide on the best course of action to take and make safe anything that is dangerous.

Put together a kit, before you have an emergency, of things you will need if the power goes down. This should comprise a wind-up torch, powder fire extinguisher, basic plumbing tools, pencil and paper, a few old cloths, a first-aid kit, a list of telephone numbers (the water supply company, neighbours, friends and emergency services), and anything else that you think might come in handy. Don't bother with a candle and matches – they are dangerous.

To DIY or not to DIY – that is the question!

Having said that DIY is a good option, there are times when it is best to call in a professional. Let's say that the water cuts off for no good reason, it's dark and cold, and you are tired and anxious. Make basic checks to ensure that everything is under control, but then turn the boiler off, shut down the main stopcock, empty the system, switch on an electrical convector heater and wait until daylight before embarking on any repair work. NEVER try to make repairs when you are tired, ill, or under the influence of drink or medication.

The eight golden rules for DIY plumbers are as follows:

- Always turn off the water at the mains and empty the system before you start cutting into pipes.
- Always pull out plugs and fuses if there is any connection between the plumbing problem and electricity – such as a problem that involves the boiler, immersion heater, and so on.

- Always be cool-headed and methodical when faced with a plumbing emergency.
- Never try DIY when you are in a rush, anxious, under pressure, feeling ill, or under the influence of drink or medication.
- Never be tempted to try a task that is beyond your skills. Call in a professional if in doubt.
- Never try to open a sealed appliance, especially if it is still under guarantee.
- Never work in the dark. Simply bring the problem to a halt and wait until the morning to remedy it.
- Always proceed on the assumption that electrical earth terminals are live and dangerous, until you have made tests with a multimeter.

Emergency task – clearing a blocked sink

The sink is full of dirty washing-up water. You pull the plug and nothing happens. Everything looks fine – there is no noise or dribbling water under the sink, and nothing unusual apart from the fact that the water doesn't go away.

TOOLS AND MATERIALS

- Torch
- Sink plunger
- Large adjustable wrench
- Old wire coathanger
- Small soft plastic pudding bowl
- Plastic bucket
- Plenty of old cloths

Method

1 Take the plastic pudding bowl, hold it rim-side-down and plunge it up and down against the waste hole. If you are lucky, you will hear a gurgling noise and the water will begin to run away.

2 If the water stays put, take the sink plunger and repeat the procedure as already described, only this time, block the overflow hole and be slightly more vigorous in your approach to it.

3 If the water stays put, use the basin to bail the water from the sink and into the bucket.

4 Remove all the items from under the sink, and put on some old workclothes and shoes.

5 With the bucket placed at the ready under the sink, take the wrench and very carefully undo the cleaning eye or, if it's a bottle trap, the lower part of the bottle.

6 Let the last dregs in the sink and the contents of the U-bend or bottle trap fall into the bucket.

7 Take the wire coathanger and reshape it so that it makes a probing hook that suits the shape of your particular trap.

8 Use the wire to fish around the U-bend or in the trap, to poke out all those long, greasy strands of hair and juicy bits of yuck.

9 Use washing-up liquid and an old toothbrush to clean up all the grease to enable the wastewater to drain away faster.

10 Reseal the trap and run water into the sink to check that it runs away. Clean up and wash your hands.

Emergency task – clearing a blocked toilet

You have just used the toilet and flushed the WC cistern, and all that has happened is that the water has topped up the WC pan to the point of overflowing. The whole pan is now full of a wonderful mixture of water, toilet paper and waste. There is a strong likelihood that the toilet is blocked with a disposable nappy.

TOOLS AND MATERIALS

- Long-handled WC plunger (from a hire shop)
- Flexible WC/drain auger (from a hire shop)
- Pair of the longest rubber gloves that you can find
- Small soft plastic throwaway bowl
- Plastic bucket
- Plenty of old cloths
- Strong-smelling disinfectant

Method

1 Dress in old clothes that can be thrown away afterwards. Put on the rubber gloves and roll back the floor covering. Spread old cloths on the floor around the WC pan.

2 If the pan is full to overflowing, you will have to use the bowl to bail out some of the contents into the bucket. Continue until the level is lower than the length of your rubber gloves.

3 Dig a hole in the garden (well away from children and pets) and bury the contents of the bucket.

4 Now for the good-fun bit. Fish around with your hands and see if you can feel a blockage in the U-bend. If you do, pull it out and, if you are lucky, you will hear a gurgling and the contents of the WC pan will drain away.

5 If you are not so lucky, and you cannot feel anything, take the long-handled pan plunger and locate the rubber head well down in the pan against the U-bend. Set to work with a vigorous push-and-pull plunging action. And once again, if you are lucky you will hear a gurgling sound and the contents of the WC pan will clear away.

6 If luck is still against you, ask a helper to go down into the garden or yard and lift the lid off the manhole inspection cover (the nearest one to the toilet window).

7 With the helper looking down into the chamber, while you remain by the WC pan, take the auger and carefully feed the flexible rod down into the first part of the U-bend. With the auger in place, slowly turn the handle until you feel a resistance, at which point the WC pan will empty.

8 The moment you see the WC pan empty, call out to your helper to observe what comes through. If tissue flows through, all well and good, but if it's something more solid such as a toy, fish it out.

9 When the job is done, wash all the equipment, the WC pan and

gloves in hot water and disinfectant. (If you are going to keep the gloves, they should only be used for similar jobs.)

10 Finally – and this is most important – tell all users of the toilet not to put anything down the toilet other than liquid, toilet tissue, and their own natural waste – no disposable nappies, sanitary towels, string, plastic bags or other items.

Emergency task – mending a slow leak in a pipe

Over a period of days, you have noticed a damp patch on a bedroom ceiling. You go up into the loft and see that the copper feed pipe running into the top of the cold-water storage tank (i.e. the rising main) has sprung a pinhole leak on the down-side of the stopcock.

TOOLS AND MATERIALS

• Copper slip coupling (15 mm) to match up with the size of the rising main

• Pair of adjustable spanners

• Pipe cutter

• Supply of old cloths

• Torch

Method

1 Go into the kitchen, pull the plug out of the kitchen sink and turn the cold-water tap on.

2 Turn off the stopcock in the cupboard under the sink.

3 Wait a few minutes for the cold-water tap to stop running, and then go up into the loft and press

down on the floating ball valve in the cold-water storage tank. If it stays silent, with no water running out, you can be certain that the leaking pipe has been drained.

4 Centre the slip coupling on the pinhole leak and use a couple of tabs of sticky tape to mark the pipe so that you know whereabouts the coupling needs to be fitted.

5 If the pipe is hardly dented at the leak, take your pipe cutter and cut through the pipe – right through the pinhole. If, on the other hand, the pipe is badly dented at the pinhole, use the cutter to take out a ring-sized slice of pipe that includes the pinhole.

6 Use a cloth to dry and clean the two ends of the pipe.

7 Ease the two pipe ends apart. Disassemble the slip coupling and slide a cap-nut, followed by an olive, on each pipe end.

8 Slide the body of the slip coupling on one pipe end, then ease the pipes back together and slide the coupling back so that both ends are contained and the body of the coupling is centred on the leak.

9 With your fingers, screw the cap-nuts as tight as possible, and then use the adjustable spanner to clench up the joints. Do not over-tighten – no more than about one-and-a-quarter turns. Wipe the coupling with dry toilet tissue.

10 Go down into the kitchen and turn the stopcock on. When the cold-water tap starts to run, turn it off. Go back to the loft and inspect the repaired joint. Allowing for

a small amount of condensation on the pipe, if the area stays dry, you can be pretty certain that the leak has been fixed properly.

Emergency task – mending a frozen burst pipe

It is late at night, freezing cold, you have just come home from a party, and you see that an outside pipe has burst – a rising main pipe. There is a long split running along the pipe.

TOOLS AND MATERIALS

- Two compression joints to fit the frozen pipe
- Generous length of pipe to match the frozen pipe
- Pipe cutter
- Two spanners to fit the joints (either two open-ended spanners, or one open-ended and one adjustable spanner)
- Warm work clothes
- Plastic bucket
- Lots of old cloths
- Hairdryer

Method

1 Go straight indoors and turn on the cold-water tap in the kitchen. Fill up a few saucepans and a couple of buckets, then turn off the outside stopcock in the yard or by the front gate.

2 When water stops running from the kitchen tap, check that the water has stopped flowing from the burst pipe.

153

3 By this time you will be both tired and cold. Make yourself a hot drink and go to bed. Use the water in the buckets to flush the toilet.

4 The next morning – after you have had a good breakfast – dress in warm clothes and take your toolkit to the burst pipe. Use the pipe cutter to cut out the section of the pipe that contains the burst.

5 Clean the cut ends and clean away all the debris. Use the hairdryer to dry off the pipes.

6 Take the two compression joints and check that they are clean and complete. Each joint should be made up from five component parts – the body of the joint, two cap-nuts and two olives.

7 One cut end of pipe at a time, slide on the cap-nut followed by the olive and the joint body. Using the distance between these two joints as a guide, cut a length of pipe to bridge the gap, so that the extra piece of pipe is a tight, sprung fit from one joint to the other.

8 Take the bridge pipe and, one end at a time, slide on a cap-nut followed by an olive.

9 Spring the bridge pipe into place between the two joint bodies. One cap-nut at a time, screw it on to the body of the joint so that the olive is captive on the pipe. Use the two open-ended spanners to clench the joints to a tight fit.

10 Finally, turn on the stopcock and close down the kitchen taps when the water starts to flow. Finally, wipe dry tissue around the pipes – just to make sure that the mend is free from leaks – clean up the mess and the job is done.

Emergency task – Clearing a blocked gully

You are emptying the kitchen sink and you hear a gurgling noise. You look at the gully outside the window – often known as the drain – and you see it overflowing and running over the yard or garden.

TOOLS AND MATERIALS

- Pair of long rubber gloves
- Workclothes
- Small throwaway container (e.g. plastic coffee mug)
- Plastic bucket
- Lots of old cloths
- Garden hose
- Disinfectant

Method

1 The gully is designed to hold and catch the solids from the kitchen sink wastewater, so this is one of those tasks that needs to be carried out every six months or so. If you eat a lot of greasy food, problems are likely to be more frequent. Put on workclothes and rubber gloves, and start by using the small container to bail out the contents of the gully into the bucket.

2 Feel around in the gully and remove as much of the gunge as you can. Be on your guard – it is always possible that you will find potentially dangerous items in the

gully, such as lengths of wire, broken glass, a fragment of razor blade or a dead rat.

3 If you live in an old house with a cast-iron gully, be very careful about the sharp edges at the bottom of the U-bend.

4 Taking the garden hose, remove the nozzle and feed the end of the hose down into the U-bend and up into the pipe run.

5 Lift the manhole cover nearest to the gully and ask a helper to go and turn the hose full on while you watch what happens.

6 Watch as the water gushes through. If you see a foreign body, such as a rubber ball or wad of plastic, do your best to catch hold of it as it enters the manhole.

7 Go back to the gully and slowly withdraw the hose.

8 Refit the nozzle and give the gully a good clean. Pay particular attention to the grid and the area surrounding the gully.

9 Clean the manhole – the metal cover, the handholds, the groove on the underside of the lid, the inside of the manhole, and so on.

10 Scrub the area around the gully thoroughly with disinfectant. Finally, remove all your work clothes, and wash your hands with soap and hot water.

Emergency task – Repairing a damaged tap

Over the weeks the cold-water tap in the kitchen has started to drip. The tap is of the old bib or pillar type. You have tried turning it off

really hard and all that has happened is that the steady drip has turned into a constant trickle.

TOOLS AND MATERIALS

• Work clothes

• A bucket

• Washing up liquid

• A screwdriver

• A small-size adjustable spanner

• A Stilson-type work wrench

• An electric paint stripper

• A roll of plastic insulation tape

• A supply of old cloths

Method

1 Let the whole household know that you are going to turn off the mains water – to allow them time to have a last drink of water, fill up the kettle, do the washing up, or use the toilet.

2 Remove the plug from the sink and turn the tap full on.

3 Fill the bucket with water.

4 Turn the stopcock off – until the flow from the tap comes to a halt.

5 With the tap still in the 'full-on' position, use your hands to undo the metal shroud that covers the upper part of the tap – in an anticlockwise direction. If it won't budge, wrap it around with the insulation tape and undo it with the Stilson wrench.

6 With the headgear nut revealed, set the small spanner on the nut and unscrew it in an anticlockwise

direction until you can lift the whole top part of the tap clear.

7 Remove the jumper part of the tap. This will be found either fixed to the underside of the bit that you have just removed, or simply sitting loose down inside the tap.

8 Locate the rubber washer. This will either be pressed onto a small knob at the bottom of the jumper, or alternatively held in place with a small nut. Remove it with either the screwdriver or the small spanner.

9 Now, either take the old washer, or – if the whole jumper unit is corroded – the whole works to a specialist supplier, and replace the unit or buy a new washer.

10 Clean the whole works in the bucket of water. Use a cloth to wipe out the inside of the tap – in order to remove any bits of old washer that remain.

11 Set the repaired/replaced jumper unit in place and make sure that it's a good fit.

12 With the washer and jumper in place, screw the headgear nut down, screw on the metal shroud, ensure the tap is fully opened, and open the stopcock.

13 Finally, very gently close the capstan head so that the water flow comes to a halt.

Emergency task – Repairing a broken toilet flush

You have a relatively modern direct-action flush toilet - one with a ceramic cistern, with most of the workings inside the cistern being made of plastic - and the flush has stopped working. Over the past week or so, it has got worse and worse - sometimes working and other times not - until now it has stopped working altogether. You can feel things happening when you pull down on the lever, but there is no flush.

TOOLS AND MATERIALS

• A good long pair of rubber gloves
• Work clothes
• A couple of metres of strong twine or an old belt
• A length of batten at about 300 mm long
• A small throwaway container – like a plastic coffee mug, or half of a tennis ball
• A plastic bucket
• Lots of old cloths
• A large adjustable wrench

Method

1 Fill the bucket up with water.

2 With the bucket of water at the ready, give all the members of the household a last chance to use the toilet - if they must - and flush it with the bucket of water. Tell them that the toilet will be out of action for a couple of hours.

3 Remove the heavy lid from the top of the cistern and put it somewhere well out of harm's way.

4 Reach into the cistern and lift the float. Bridge the top of the cistern

with the wood and tie the float arm up so that the water stops running.

5 Use the bowl/cup to bale the water out into the toilet pan.

6 With the cistern more or less empty, take the spanner/wrench and undo the large plastic nut on the underside of the cistern. Move the outlet pipe to one side.

7 Depending on your model of cistern, undo the nut or clamps, disconnect the wire that links the flushing lever to the top of the siphon, and then lift the whole siphon clear of the cistern.

8 You have a choice at this stage. You can either replace the whole siphon unit, or you can remove the pierced plastic plate from the underside of the siphon, and remove and replace the plastic flap valve. Whichever you choose to do, it's a very simple operation.

9 Finally, when you have fitted a new flap valve, or purchased a whole new unit, you simple reverse all the fitting procedures as already described, and the job is done. Note - the trick with this task is to make sure that you thoroughly clean and dry all the washers, nuts, and surfaces prior to refitting.

Emergency task – Repairing an overflow

You see that there is water running from the overflow pipe - the one that comes out from the wall or from under the roof eaves. It started as a small drip and now it is a steady torrent. You have been up into the loft and seen that if you leave it much longer the overflow won't be able to keep up with the rise of water in the tank - to the point where the tank might brim over.

TOOLS AND MATERIALS

- A good long pair of rubber gloves
- Work clothes
- A couple of metres of strong twine or an old belt
- A Stilson wrench
- A pair of pliers
- A pair of mole grips
- A screwdriver
- Fine grade wire wool
- Vegetable oil
- Torch

Method

1 First of all go up into the attic and have a good look at the cold water storage tank. Almost certainly you will see that the water level is slightly higher than the overflow pipe. If this is the case – and it usually is – then the problem is with the ballcock float valve.

2 Set out all your tools so that they are comfortably to hand. If there is no mains light in the attic, arrange the torch so that it's shining on the float valve. Better still; see if you can get a friend to help. WARNING: On no account must you have a mains table lamp or similar balanced over the tank.

3 Turn the water off at the mains.

4 Go back up to the attic and check

157

that the water has stopped flowing into the tank. Drain off some of the water so that the water level is much lower than the valve.

5 Have a closer look at the valve. If it's a Portsmouth valve - with the piston travelling in a horizontal direction - take a pair of pliers and remove the split-pin so that the float arm drops free. Be careful that you don't lose anything into the water. Note: if the valve is an older pattern, the piston will run vertically rather than horizontally, but other than that, the following procedures will be much the same.

6 Take a pair of grips to the cap at the end of the body of the valve and unscrew it by turning it in an anticlockwise direction. Go at it gently, all the while being careful not to twist the valve or crack the cap.

7 With the cap out of the way, gently insert the screwdriver into the underside of the piston tube – where the float arm was fitted – and carefully ease the piston out from its chamber.

8 Take the piston to the light so that you can see what's going on. Depending upon the design, the washer will either be contained by an open-ended cap, or simply set in a recess. So… either use the grips and the screwdriver to unscrew the cap that contains the washer, or simply use the screwdriver to dig the washer out from the recess.

9 Take the fine-grade wire wool and the vegetable oil and clean the whole valve inside and out. Pay particular attention to the outside face of the piston and the inside face of the piston chamber.

10 Finally, replace the washer, reverse all the procedures as described and turn the water back on.

Emergency task – Repairing a partially cold radiator

You are sitting in one or other of your rooms, with the central heating full on, and yet it feels a bit cold. You feel the radiator and observe that the top half is cold. All the other radiators in the house are hot. You realize that there is air trapped inside the radiator and that it needs 'bleeding'.

TOOLS AND MATERIALS

- Work clothes
- A pair or pliers
- A pair of mole grips
- A radiator key to fit your radiator
- Plastic sheet
- Masking tape
- Lots of cloths and newspapers
- A throwaway soft plastic container
- A pair of scissors
- A small knife

Method

1 Have a close look at the radiator and find out which of the two top corners contains the bleed valve.

2 Prepare the area around the radiator – tying back curtains, moving furniture out of the way,

rolling back carpets and so on.
If the carpet is fitted, then cover
the area around the radiator with
newspapers and old cloths. Slide a
sheet of plastic behind the corner
of the radiator so that the wall is
protected, and fix it in place with
the masking tape. Arrange all your
tools so that they are conveniently
to hand.

3 Switch off the boiler.
4 Take the scissors and trim the soft
plastic container so that it nicely
cups the valve at the top of the
radiator. Spend time making sure
that it is a good fit.
5 If your radiator has been heavily
painted over the years – and most
have – the bleed valve will be
bunged up with hard paint. If this
is the case, use the knife to scrape
the paint away from the valve area.
WARNING: Before you start this
procedure, take note that you only
need to open the valve by about a
QUARTER TURN – just enough to
let the air out. You can't speed the
operation up by removing the
valve. Work in silence so that you
can hear the hiss as the air escapes.
6 With the plastic container held
and cupped in one hand, the key
held in the other, and with your
ear close by, fit the key in the
valve and very gently turn it in
an anticlockwise direction – until
you hear the hiss as air escapes.
7 With the key still held tight in
the valve, continue listening and
watching, until the hissing stops
and the first drop of water appears.
When this happens, tighten the key.

8 Finally, turn the heating back on
and tidy up any mess.

Emergency task – Draining the system

There is a plumbing fault with your
central heating system that needs to
be fixed, to the extent that you have
to drain the system.

TOOLS AND MATERIALS

• A good long pair of rubber gloves
• Work clothes
• Lots of old cloths
• A hose pipe long enough to run from the drain cock near the boiler to the nearest door to the garden/yard/drive
• A batten long enough to bridge the feed-and-expansion tank
• A couple of metres of strong twine, or an old belt
• A pair or pliers
• A pair of mole grips
• Screwdriver
• Torch

Method

1 Turn the boiler off so that the
pump is not operating.
2 Wait until the water starts to cool.
3 Turn off the main stopcock to the
supply feeding the expansion tank.
4 Go up into the attic and, as a
precaution – and just in case the
stopcock is faulty or you have
turned off the wrong stopcock –
bridge the tank with the batten
and tie up the float valve arm so
that the water stops running.

5   Go down to the drain cock. Slide one end of the hose onto the drain cock outlet and run the other end through the house to the nearest door and out to one or other of the garden drains. Tie the hose securely so that it can't fall off the drain cock nozzle. NOTE: The hose must run level, not uphill. Keep children and pets out of harm's way.

6   With a supply of old cloths around the drain cock, use a spanner or key to open up the valve.

7   Keep a watchful eye on both ends of the hosepipe – especially the end on the drain cock.

8   When the water stops running from the end of the hose, go from one radiator to another slightly opening the bleed valves, so that the last dregs of water run out.

9   When you have made your repairs, close the drain cock and the radiator bleed valves, remove the stick and string from the tank, and turn on the main stopcock.

10 Working from the topmost radiator, bleed each radiator in turn – as described elsewhere in the book. Finally, bleed the water pump and tidy up the mess.

Emergency task – Replacing a damaged radiator

You see that there is a pool of water under one or other of the radiators. On close inspection you see that there is damage to this particular radiator – none of the others – to the extent that it alone needs to be replaced.

TOOLS AND MATERIALS

- A new radiator of the same size, type and design as the damaged one
- Work clothes
- Lots of old cloths
- A bucket
- A plastic bowl low enough in height to fit under the radiator valve
- A Stilson wrench
- A pair or pliers
- A pair of mole grips
- Screwdriver
- Permanent felt tip marker

Method

1   Roll back the carpet around the radiator or cover the area with plastic, newspaper and old cloths.

2   Identify the two valves – the hand wheel valve and the lock shield valve. Use the felt tip marker to draw alignment marks on the lock shield valve – on the spindle and the body of the valve.

3   Turn the hand wheel valve to the 'off' position. Repeat with the lock shield valve, using the felt tip marks to count the number of turns – so that you can re-open the valve by exactly the same amount.

4   Use the bleed key to open the bleed valve.

5   Go to one or other of the valves and use the spanners to loosen the cap nut – so that it is finger loose.

6   With the bucket close at hand, set the bowl under the loosened valve and continue undoing until a

dribble of water flows into the bowl. When the bowl is almost full, tighten the cap nut slightly and transfer the water to the bucket.

7 Continue decanting the water from radiator, to bowl, to bucket, until the water stops flowing.

8 Empty the bucket of water.

9 Unscrew the cap nut from the valve at the other end of the radiator.

10 Carefully lift the radiator from its brackets, and just as carefully empty the last dregs of water from the radiator into the bucket.

11 Take the radiator out of the room.

12 Hang the new radiator on the brackets, and use the spanners to tighten the cap nuts on the two valves – the lock shield valve and the hand wheel valve.

13 Having first made sure that the bleed valve is closed, open the hand wheel valve to the 'full on' position. Then open the lock shield valve by the same number of turns that it took to close it.

14 Finally, having checked that both valve connections are tight and dry, gently open and close the bleed valve until the trapped air has been released, and the job is done.

Emergency task – Curing an airlock in the kitchen hot water tap

No water is coming out of the hot tap in the kitchen. You have drained the system, drawn off huge amounts of water, shut down the stopcock or whatever, with the effect that while the hot tap makes lots of grumbling noises, it doesn't work. When this happens, it usually means that there is a pocket of air trapped in the system.

TOOLS AND MATERIALS

- Work clothes
- A bicycle pump
- A length of garden hose – long enough to loop from the hot tap to the cold
- A couple of tap adaptors to fit the hose

Method

1 First try out the bicycle pump.

2 With the faulty tap fully open, stick the pump adaptor up the spout of the tap and seal around it with wet cloths, while a friend works away at the pump. The tap may make noises and start running.

3 If the pump method doesn't work, fit an adaptor on each end of the length of garden hose.

4 Fit one end of the hose on the offending tap and the other end on the cold-water tap.

5 With the faulty tap fully open, turn the cold water tap full on.

6 After a few seconds, the pressure of water from the rising main will blow out the bubbles of trapped air, with the effect that the tap will be back in service.

7 NOTE: If the air lock keeps occurring, either the main supply tank is too small and empties faster than it fills – in which case you need to fit a larger tank – or the ball valve is faulty and needs to be repaired or replaced.

Glossary of tools and techniques

Adjustable spanner
A spanner with a screw-operated lower jaw. In use, you open up the jaws, slide the spanner over the nut and tighten up for a good, firm fit. An adjustable spanner is an essential tool. Buy the best that you can afford, and get several sizes.

Aligning
The procedure of setting one part on or against another – perhaps the sink on a wall – with the help of a measure and spirit level in order to obtain a good fit.

Basin wrench
A long-handled wrench designed specifically for use when fitting taps. In use, the jaw of the wrench is flipped over (so that it is in either the tightening or untightening mode) and located on the tap connector, and then the long handle is operated with a pivoting movement. This is an essential tool if you want to fit a tap – especially if the tap is being fitted on a small bathroom basin.

Bending spring
A spring used for bending small-gauge pipes (22 mm or less in diameter). In action, you slide the bending spring into the pipe, and then bend the pipe over your knee. The spring prevents the pipe from kinking at the bend. The spring is then removed with a twisting action. The bending spring is an essential tool if your project is one that will necessitate bending pipe.

Chain wrench
A tool with a handle, a toothed jaw and a chain – used for gripping pipes and large nuts. In action, you wrap the chain over the item to be turned, engage the chain with the toothed jaw, and then lever the handle so that the chain grips. The chain wrench is a very useful tool, especially if you have to work on ancient pipework, large nuts, threaded galvanized pipe and suchlike. Ideally, you need a chain wrench and a Stilson wrench.

Cold chisel
A tool used for general cutting (such as hacking old pipes or plaster), which is used in conjunction with a club hammer. In action, you hold the chisel with one hand and give it a well-aimed blow with the hammer. Always wear gloves and goggles. An essential tool. Note that some cheap, bargain chisels (usually made in China) are so soft that they are practically useless. Buy a best-quality British or American tool, either new or secondhand.

Craft knife
Just about any sharp-bladed knife of the 'Stanley' type – used for all general cutting tasks.

Cranked spanner
A tool used for working on tap connectors. It can be used alongside or instead of a basin wrench. A cranked spanner is an essential tool if you plan to work on new or existing taps. You will need either or both tools.

Drain rods

A set of flexible, screw-ended rods used for clearing blockages in drains and manholes. In action, you enter the first rod into the drain, complete with the chosen head (plunger, scraper etc.), screw the next rod on to the first, and so on, all the while pushing and turning the rod in a clockwise direction. This is an essential piece of equipment for the homeowner, especially if you live in the country and have a cesspit or septic tank. Always pull and push with a clockwise action, otherwise the sections will unscrew.

Drills and drilling

You need two drills, an electric drill for heavy-duty drilling, such as making deep holes in brickwork, and a hand drill or cordless drill for use when the power has been switched off. Stock up with a good selection of drill bits in a range of diameters and lengths, including a masonry bit about 250 mm long (long enough to drill through a brick wall) and of a diameter large enough to take the pipe in question.

Engineer's vice

A large, metal vice used for gripping pipes and suchlike. In use, the vice is bolted or clamped to a bench, the workpiece is set in its jaws and held securely, and the rod handle is turned. This is a very useful piece of equipment – not essential, but it does make some tasks a lot easier. A good secondhand vice can often be purchased at a car-boot sale, but it is best to buy new – especially if you are planning a lot of plumbing work.

Files

Files are used for all the little smoothing, rubbing, cleaning and shaping tasks involved in plumbing. In use, the file is gripped by the handle, the working face or edge of the file is brought into contact with the workpiece, and the file is moved in a backwards and forwards action – with the cutting being done on the forward stroke. We use a set of a dozen files, ranging in profile from flat to round, half-round, triangular and square. The sizes vary from very large (about 300 mm long and 60 mm wide), through to a set of needle files that are not much bigger than a slender penknife. Files are essential tools. WARNING: Metal filings are potentially very dangerous – very nasty if they get in your eyes, or become embedded in your hands or feet. Make a special point of cleaning them up, especially if you have children or pets.

Finishing

The procedure of tightening screws, wiping filler into holes, grouting tiles, cleaning up and so on, necessary in order to bring a plumbing job to a happy conclusion.

Fireproof mat

A square of fireproof material used when soldering to protect surrounding features. In action, you place the mat between the joint being worked and, for example, the wall. This is an essential piece of equipment, especially if you need to work on existing pipework.

Floorboard chisel (bolster chisel)
Sometimes called a bolster chisel, this wide-bladed cold chisel is used in conjunction with a club hammer to lift floorboards and to chop holes in plaster and brickwork. This is an essential tool if you need to put pipes under a wooden floor.

Gas blowtorch for soldering
A small, hand-held blowtorch that runs on liquid gas is an essential tool used for soldering joints. Properly handled, it is safe, clean and completely easy to use – perfect for beginners and professionals alike. In action, you simply play the torch over the joint until you see the solder flow. When the gas runs out you fit a new cannister. If you are a beginner, practise your technique with a bag of pre-soldered joints and bits of copper tube. Yes, you could always use an electric soldering iron – but it is not easy.

Hacksaw
A saw used for cutting metal. You will need a couple of hacksaws – a large one for general tasks such as cutting bolts, pipes and wires; and a small, 'junior' hacksaw for working in confined spaces. Make sure that you have a good supply of spare blades, especially if you plan to work in the evening or over a holiday period. An essential tool.

Hammers
We use three hammers – a claw hammer for general work (banging in nails and clawing out old screws), a small tack hammer (for banging in pins and small nails), and a large-headed club or lump hammer (used in conjunction with a flooring or bolster chisel, to lift floorboards and chop holes in plaster and brickwork). Hammers are essential tools.

Hydraulic pump and sink plunger
A small push-and-pull tool used for clearing blocked waste pipes and sinks. In use, you put the rubber cup in place over the outlet and push and pull the handle to create a powerful jet of water. This tool costs more than a sink plunger, but it is both faster and more efficient. A hydraulic pump or a sink plunger is an essential tool, especially if you live in the country and have a cesspit or septic tank.

'Junior' hacksaw – *see* **Hacksaw**

Measuring tools
You need two measuring tools – a wood, plastic or metal measuring rule for sizing and marking things such as the precise size and position of holes, and a flexible tape measure for general measuring – establishing the position of taps, radiators etc., and for measuring lengths of pipe.

Multimeter
Sometimes referred to as a continuity tester, this is a low-cost piece of battery-operated equipment used for testing the flow of electric current through cables, bulbs, fuses and suchlike. Although a multimeter is essentially an electrical tool, it is very useful when you are fitting earth wires to pipes and testing boilers.

Pliers
A good pair of pliers is very useful. We use long-nosed pliers for working with electrics, a large pair of pliers for heavy work such as soldering and adjusting WC valves, and a really huge pair for all manner of 'can't-quite-reach-it' tasks.

Plier wrench
Popularly known as a mole wrench, this looks like a pair of pliers with a nut-and-lever adjustment. In use, the nut is loosened, the jaws are located on the workpiece, the nut is tightened, and then the lever is adjusted until the handles can be snapped

together. It isn't an essential tool, but is amazingly useful – for gripping a hot pipe, holding soldering wire, gripping tin plate, undoing awkward nuts, and so on. We have three or four different sizes and they all come in useful.

Ring spanner
A long-handled wrench with a closed ring head at either end. If you can slide the ring over the end of the pipe, this is a good tool for removing corroded cap-nuts from existing pipe runs.

Rotary pipe cutter (clamp type)
A small, hand-held tool used to cut copper pipe. In action, you sit the tool on the pipe (like a saddle on a horse), tighten up the clamping screw, and then turn the tool around the pipe. As you turn, you tighten up the screw until the pipe has been cut. The joy of this tool is that it makes a perfectly clean 90° cut and leaves the end of the pipe feeling smooth – there is no need to clean it up with a file. This is an

essential tool, especially if you need to work on existing pipe runs.

Rotary pipe cutter (ring type)
A small, hand-held tool (also known as a pipe slice) used for cutting copper pipe. In use, you slip the little disc over the pipe and turn it until the cut has been made. It isn't an essential tool, but is a great timesaver when you need to cut into an existing pipe run that is fitted close to the wall or in a confined space.

Saws
Generally, you need a saw to cut through floorboards and plasterboard walls in readiness for running pipes, and a saw for cutting wood. We use two saws: a traditional, short, curve-ended handsaw designed specifically for cutting floorboards, and an electric jigsaw. The jigsaw is particularly good for cutting holes in floors and plasterboard. In use, you drill a starter hole, enter the blade into the hole so that the bed of the tool is resting flat

on the floor/wood/board, switch on the power, and then slowly advance the tool to cut to the waste side of the drawn line. Switch off the power, wait for the blade to come to a standstill and then lift the tool away. Make sure you know there are no cables or pipes hidden behind or beneath the material you are cutting. WARNING: Never lift the tool away from the workpiece while the blade is still in motion, and always wear goggles and a facemask.

Screwdrivers
You need a good range of screwdrivers in a variety of sizes. We use everything from general-purpose screwdrivers for driving in wood and masonry screws, through to special electricians' screwdrivers. When choosing electricians' screwdrivers, always get the best that you can afford, and choose those that are insulated along most of their length, with the insulation covering the whole of the handle and the metal shaft, stopping just short of the

end of the blade. Make sure that at least one of your small screwdrivers has a magnetic tip so that you can retrieve screws and nuts from otherwise inaccessible cavities.

Sink plunger
A tool used for clearing blocked sinks and waste pipes. It has a wooden handle on one end and a rubber cup on the other. In use, you fill the sink up with water, position the rubber cup so that it covers the waste hole, and then set to work pumping. Rubber plungers are cheap and can last forever, but are not so efficient as hydraulic pumps. It is essential to have one or other of these tools.

Spanners
Traditional open-ended spanners are really good tools for working on compression fittings. If you are going to work on an old system – where the pipes are variously heavily painted, in Imperial sizes or made from galvanized steel – it is a good idea to get a range of secondhand

spanners in Imperial and metric sizes. Our spanners come from car-boot sales.

Spirit level
A tool used for checking that fittings, fixtures and components are vertically and horizontally level or true. Spirit levels come in all manner of types and sizes. We prefer to use a small wood and brass level for checking bathroom fittings, and a cheap, throwaway, plastic level for checking things such as pipe runs.

Stilson wrench
A long-handled, tooth-jawed adjustable wrench used for gripping cap-nuts and round sections. In action, the wrench jaws are opened so that they just slip over the item that you want to turn or undo, and then the handle is pulled so that the levering action results in the jaws biting and gripping the workpiece. This is an essential tool if you intend to work on an ancient plumbing system. It is a good-quality tool

that is virtually indestructible. We use a Stilson wrench that has been handed down through the family over the generations and must be at least 80 years old – and it seems to get better every time we use it.

Tin snips or metal snips

A scissor-like tool used for cutting thin metal sheet. In action, you hold the workpiece in one gloved hand while operating the snips with the other. If possible, choose 'universal' snips, which have thick, narrow blades. They are stronger and more versatile.

Torch

A torch is a must-have piece of equipment – just the thing for when the power has been switched off and you are crawling about in the loft or under the floor. You need a top-quality torch that can be both held or stood on its base. We use a wind-up model so that we never have to worry about batteries. WARNING: In the context of working in the attic on the water storage tank, never, in the absence of a torch, be tempted to rig up a temporary bulb-and-flex lamp. While this would certainly give you a better light, there's a danger that you could be fiddling about over the tank, with wet hands, and the light could fall into the water. The results could be fatal.

Wire strippers

A small tool, a bit like a pair of pliers, used for stripping plastic insulation from cables and flexes. Good when fitting earth bonding cables to pipes.

Wood chisels

You need a set of chisels for generally chopping and cutting wood. We use a set of low-cost chisels for this, and for cutting holes in plasterboard. In use, you hold and guide the chisel with one hand, and tap it with a hammer or mallet held in the other hand. Always wear goggles and a mask when using a chisel.

Glossary of terms, tips and materials

Aligning
The procedure of setting one part on or against another, such as a socket on a wall, with the help of a measure and spirit level in order to obtain a good line and fit.

Airlock
An airlock is a trapped pocket of air in a pipe, which cuts off the water supply to a tap. A possible cure is to get a length of garden hose, slip it over the affected tap and the kitchen cold-water tap, and to turn both taps full on. With a bit of luck, the mains-pressure water running from the cold-water tap will force out the pocket of air. If this doesn't work, try slipping a short length of hose on the affected tap and blowing down it with either a bicycle pump or a car tyre pump.

Burr
A rough, raised, serrated edge on the cut end of a pipe, left by using a hacksaw to make the cut. Such a burr can be removed with a fine-toothed file and/or a piece of wire wool. A pipe cutter will cut a pipe without leaving a burr.

Cable
DIY plumbers need to know something about electric cables. A cable is a device that provides a path along which the electricity flows. It is made up of insulated metal conductors, or cores, and covered with a protective outer sheath of plastic or PVC. Most cable has three cores, the live sheathed in red, the neutral in black and the earth bare. The bare earth between the end of the cable and the fitting it is connected to, must always be protected by green/yellow earth sleeving. Cable comes in various sizes and ratings, each designed for a specific task. If in doubt of the type to use, consult a specialist supplier, detailing the work planned, and then buy the cable by the roll.

Cap-nut
The nut part of a compression joint. In action, the cap-nut squeezes the olive into the body of the fitting, with the result that the olive bites and grips the pipe being joined.

Cesspit
A large, buried tank used to collect sewage and wastewater. Cesspits are a bad option in that they have to be emptied at regular intervals. *See also* **Septic tank**.

Cistern
Another name for a water storage tank. For example, both the tank behind the WC and the cold-water storage tank in the attic can be called cisterns.

Consumer unit
A unit, sometimes known as the main fuse box, often found in the garage, under the stairs or in its own special cupboard, which contains all the switches, fuses and/or miniature circuit breakers (MCBs) governing the different circuits that run around your house. In older houses, the fuse box was/is usually a rather large black, brown or cream box, in which

each of the circuits is protected by a rewireable fuse. When there is a problem, the fuse burns out and you have to turn OFF the main power supply, identify and pull out the offending fuse block, replace the fuse wire, solve the problem that caused the fuse to blow and then switch the power back on. With a modern installation, the fuse box is replaced with a neat white box complete with a row of small switches or miniature circuit breakers (MCBs). When there is a power cut, all you have to do is turn OFF the main power supply, identify and sort out the problem, switch on or push back the MCB, and then turn the power back on.

Cross-bonding
Cross-bonding is a system to protect the occupants of the house from electric shock in the event of a short-circuit. The various high-risk metal items in your home – such as the boiler, stainless-steel sink, metal tank and metal bath –

are protected by a dedicated earth cable that links them directly to the earth terminal in the fuse box or consumer unit. In such a system, the earth wire might run from the copper plumbing pipes to the cast-iron bath, then on to the stainless-steel sink, the stove in the kitchen, the metal water tank in the loft, and so on, finishing up at the consumer unit. The earth cable is attached to each item with a clamp or clip that includes a metal tag. If there is a problem, the dangerous renegade currents are swiftly and safely run back along the dedicated earth cable to the fuse box/ consumer unit.

Draincock
The draincock is the lowest tap in an appliance (such as a boiler or radiator) or system, from which water can be drained. To drain a system, you simply slip a garden hose on the draincock, run the other end of the hose out into the garden and then turn on the tap.

Earthing
A safety feature that protects the user by supplying an easy, safe passage for renegade currents, which would otherwise flow into a person or appliance. Every circuit supplies this passage via the green/yellow conductor that runs alongside the live and neutral. The short length of bare earth core that runs from the end of cable or flex to a switch, socket, or appliance must be protected by a length of green/yellow earth sleeving. Large high-risk metal items, like iron baths and steel sinks, must be connected to a dedicated earth that runs back to the consumer unit or fuse box. *See also* **Cross-bonding**.

Finishing
The procedure of tightening screws, adjusting the level of fittings, sweeping up, wiping filler into holes etc. in order to bring the job to a clean and tidy conclusion.

Float valve
A float valve, such as the

169

one in the WC cistern, is operated by the movement of the water in the tank. In action, the flush is operated, the water level goes down, the ball float falls, and the valve is opened. And of course when the water level rises, the hollow ball floats, and the water gets cut off.

Fuse box –
see **Consumer unit**

Fuse
A length of wire that is designed to melt at a certain current demand and in so doing cut the circuit. Fuses are colour coded for easy recognition: white = 5 amp, blue = 15 amp, yellow = 20 amp, red = 30 amp and green = 45 amp. Always replace a fuse with one of the same rating and never use anything other than the recommended fuse wire or cartridge.

Fused connection unit
A unit that allows a junction between a higher-rated cable and a lower-rated cable or flex, with a fuse that will blow

to prevent the smaller cable being overloaded.

Hopper head
The bucket-shaped container at the top end of a rainwater pipe, installed when a gully or a number of waste pipes need to run into the same drainpipe. The hopper acts like a funnel.

Insulation tape
Plastic PVC sticky-back tape comes in a range of colours to match the colour-coded sheathing on electrical cables and flexes. It is used to identify and insulate cores on switch circuits, where the bare core runs from the cable to the terminal; it can also be used to label pipe runs, bind up low-pressure leaks, and so on.

Miniature circuit breaker
The little switches, also called MCBs, in modern consumer units that fulfill the same function as a fuse. If there is a problem, the appropriate MCB trips and cuts off the power. Simpler than working with a fuse, you

solve the problem and then turn the MCB switch back on.

Overflow pipes
An overflow pipe is fitted to tanks and cisterns at the 'maximum height'. If there is a problem such as a faulty float valve, the rising water safely runs down the overflow, rather than doing damage by brimming over the top of the tank.

PTFE tape
A thin, ribbon-like plastic material used to seal threaded fittings. In action, the tape is wrapped around the thread and the joint is tightened up so that the tape fills up the 'V' section part of the thread. Many plumbers prefer to use the traditional method of wiping the thread with a mastic made from putty, and then finishing by winding over a thread or two of hemp fibre.

Residual current detector
A safety device, also known as an RCD or

residual current breaker (RCB), fitted as a permanent fixture in a circuit or as a portable item with a socket to protect you from renegade currents. If there is a problem, the device cuts the power off instantly, fast enough to protect you from harm.

Rising main
The cold-water pipe that runs from the water supplier's stopcock to feed water (under pressure) to the kitchen tap and cold-water storage tank in the loft. The first thing you do in a plumbing emergency is to turn off the stopcocks to cut off the rising main.

Septic tank
A buried tank system that both stores and treats sewage and wastewater. A good septic tank is self-contained and hardly ever needs emptying. In use, the waste runs out of the house through an underground pipe, through one or more inspection chambers, and on into a large holding tank. Over time, the heavy part of the waste falls to the bottom of the tank as sludge, and the relatively clear water that remains runs off through one or more filter beds. It takes quite a few years for the sludge at the bottom of the tank to rise to a level that will necessitate its being emptied. REMEMBER: A cesspit is a constant nuisance as it frequently needs to be emptied and can be smelly, whilst a septic tank in good condition is relatively trouble-free, doesn't smell offensive and will hardly ever need to be emptied. *See also* **Cesspit.**

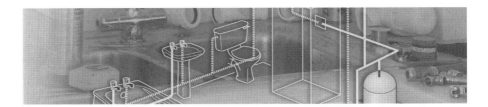

Useful contacts

Associations

**British Plumbing
Employers Council
(Training) Ltd**
2 Mallard Way
Pride Park
Derby
DE24 8GX
Tel: 0845 644 6558
Fax: 0845 121 1931
Web: www.bpec.org.uk

CORGI
Tel: 0800 915 0480
Web: www.trustcorgi.com

Energy Savings Trust
Tel: 0800 512 012
Web: www.est.org.uk

Environment Agency
Tel: 08708 506 506
Web: www.environment-
agency.gov.uk

**Heating and Hotwater
Industry Council**
36 Holly Walk
Leamington Spa
Warwickshire
CV32 4LY
Tel: 0845 600 2200
Fax: 01926 423 284
Web: www.centralheating.
co.uk

**Institute of Domestic
Heating &
Environmental
Engineers (IDHEE)**
Dorchester House
Wimblestraw Road
Berinsfield
Wallingford
OX10 7LZ
Tel: 01865 343 096
Fax: 01865 340 181
Web: www.idhee.org.uk

**National Home
Improvement Council**
Roofing House
31 Worship Street
London
EC2A 2DY
Tel: 020 7448 3853
Fax: 020 7256 2125
Web: www.nhic.org.uk

SGAS
Specialist Gas Assessment
Services Ltd
8 Broughton Way
Off Thompson Road
Whitehills Business Park
Blackpool
FY4 5PN
Tel:01253 697 078
Fax: 01253 768 162
Web: www.sgas.co.uk

**The Institute
of Plumbing and
Heating Engineering**
64 Station Lane
Hornchurch
Essex
RM12 6NB
Tel: 01708 472 791
Fax: 01708 448 987
Web: www.iphe.org.uk

**Copper Development
Association UK**
5 Grovelands Business
Centre
Boundary Way
Hemel Hempstead
Herts
HP2 7TE
Tel: 01442 275 700
Fax: 01442 275 716
Email: copperboard@
copperdev.co.uk

Tools and materials suppliers

Armitage Shanks
Tel: 01543 490 253
Fax:01543 491 677
Web: www.armitage-
shanks.co.uk

B & Q plc
Head Office
Portswood House
1 Hampshire Corporate
 Park
Chandlers Ford
Eastleigh
Hants
SO53 3YX
Tel: 0845 609 6688
Web: www.diy.com

BaxiPotterton Limited
Brownedge Road
Bamber Bridge
Preston
PR5 6SN
Tel: 08706 060 780
Fax: 01772 695 410
Email: info@baxi.co.uk

Black and Decker
210 Bath Road
Slough
Berks
SL1 3YD
01753 567 055
Web: www.blackanddecker.
 co.uk

Bristan Group Ltd
Birch Coppice Business
Park
Dordon
Tamworth
Staffordshire
B78 1SG
Tel: 0870 4425556
Web: www.bristan.com

Focus (DIY) Ltd
Head Office
Gawsworth House
Westmere Drive
Crewe
Cheshire
CW1 6XB
Tel: 0800 436 436
Web: www.focusdiy.co.uk

Homebase Ltd
*Use the online store
finder for local contact
information.*
Web: www.homebase.
 co.uk

Honeywell
2480 Regents Court
The Crescent
Birmingham Business Park
Birmingham
West Midlands
B37 7YE
Tel: 0121 480 5200
www.honeywell.com

**Polypipe Building
Products Ltd**
Broomhouse Lane
Edlington
Doncaster
DN12 1ES
Tel: 01709 770 000
Fax: 01709 770 001
Web: www.polypipe.com

Stanley UK Ltd
Stanley UK Services Ltd
Europa View
Sheffield Business Park
Sheffield
S9 1XH
Tel: 0114 244 8883
Web: www.stanleyworks.
 co.uk

Screwfix
*Local trade counters
throughout the UK.*
Tel: 0500 41 41 41
www.screwfix.com

Tilgear
Bridge House
69 Station Rd
Cuffley
Herts
EN6 4TG
Tel: 01707 873 434

Vaillant Ltd.
Vaillant House
Trident Close
Medway City Estate
Rochester
Kent
ME2 4EZ
Web: www.vaillant.co.uk
Tel: 01634 292300
Fax: 01634 290166

Index